Working at Inventing:

Thomas A. Edison and the Menlo Park Experience

Edited by
William S. Pretzer

The Johns Hopkins University Press
Baltimore and London

Originally published by Henry Ford Museum & Greenfield Village, 1989
Johns Hopkins Paperbacks edition, 2002
2 4 6 8 9 7 5 3 1

Photographic research by John Tobin
New photography by Rudy T. Ruzicska, Alan R. Harvey, and Tim Hunter

The Johns Hopkins University Press
2715 North Charles Street
Baltimore, Maryland 21218-4363
www.press.jhu.edu

Library of Congress Cataloging-in-Publication Data

Working at inventing : Thomas A. Edison and the Menlo Park experience /
edited by William S. Pretzer.—Johns Hopkins paperbacks ed.
p.   cm.
Originally published: Dearborn, Mich. : Henry Ford Museum &
Greenfield Village, 1989.
Includes bibliographical references and index.
ISBN 0-8018-6890-4 (pbk. : alk. paper)
1. Menlo Park Laboratory. 2. Edison, Thomas A. (Thomas Alva),
1847–1931. I. Pretzer, William S.
T178.M46 W67 2002
621.3'092—dc21   2001037744

A catalog record for this book is available from the British Library.

Contents

Preface to the Johns Hopkins Edition

THIS BOOK WAS CONCEIVED IN 1989 as an accessible antidote to the "heroic inventor" image that still pervades the public mythology of Thomas Edison and, indeed, most inventors. The inspiration for the book came from the research conducted as the staff of Henry Ford Museum & Greenfield Village renovated the Menlo Park buildings, which were first reconstructed in Henry Ford's outdoor history museum in 1929. More specifically, the book evolved from questions public historians ask themselves all the time: What lessons or morals will members of the general public take away from our presentation? What is the most "usable past" we can provide for the everyday citizen?

Scholars debate intellectual theories and provide ever more precise explanations of historical nuance. Students (from elementary school kids to graduate students) must master what their mentors (teachers and state boards of education as well as Ph.D. advisors) decree as mandatory. But what about people without a learning agenda, without deep historical knowledge or even the desire to question the history they vaguely remember? What of the Detroiters who want to show their out-of-town relatives a local cultural attraction? What is memorable for the children who get dragged where their parents want to go, let alone for the active retirees who go where they want? What can a history museum offer them?

In the case of Thomas Edison and Menlo Park, the project team decided that two overarching, useful lessons could be taught through this historic site. First, it seemed worthwhile to show how much Edison relied on others, how interdependent he was with the team of workers who shared his life and work in Menlo Park, New Jersey, from 1876 to 1882. This lesson cuts against the grain of the solitary inventor myth and resonates with experiences common today. Second, the lesson that persistent, rigorous work—as opposed to the flash of insight—is integral to the inventing process seems to provide value to people. As teachers, we seldom can inspire individuals to personal "genius," but we can inspire them to value collaboration and recognize the hard work of creativity. "Opportunity," said Edison, "is missed by most people because it is dressed in overalls and looks like work."

Of such lessons were born the public interpretation at the Menlo Park site in Greenfield Village and this publication, *Working at Inventing*. Both of these presentations emphasize the importance of the organization, communication, and application of diverse skills. A little over a decade later, nearly 6 million people (including 2 million school children) have been introduced to these ideas through a visit to the Menlo Park complex, and more than 5,000 copies of this book have been purchased.

Working at Inventing has found a place on the bookshelves of professional and amateur historians alike who take Edison as their subject or their muse. Topics and perspectives offered by the authors of these essays have influenced subsequent writers; the four biographies of Edison published in the 1990s have relied on material from this book.* The contributors to *Working at Inventing* never intended to have the last word on those six frenetic years at Menlo Park that gave birth to the era of modern sound and light technologies, and by reissuing this book, the Johns Hopkins University Press virtually ensures that these authors' insights will be not be lost.

William S. Pretzer, Curator
Henry Ford Museum & Greenfield Village

---

*Martin V. Melosi, *Thomas A. Edison and the Modernization of America* (Glenview, Ill.: Scott, Foresman/Little Brown, 1990); Neil Baldwin, *Edison: Inventing the Century* (New York: Hyperion, 1995); Gene Adair, *Thomas Alva Edison: Inventing the Electric Age* (New York: Oxford University Press, 1996); and Paul Israel, *Edison: A Life of Invention* (New York: John Wiley, 1998).

Preface to the Original Edition

INNOVATION, INGENUITY, AND ENTERPRISE at Thomas Alva Edison's Menlo Park Laboratory complex is the subject of this volume. The theme is especially appropriate to Henry Ford Museum & Greenfield Village. This museum complex, which occupies 260 acres in Dearborn, Michigan, was founded by Henry Ford in 1929 as a living memorial to these concepts as embodied in the life of Thomas Alva Edison. The official corporate name of the museum complex is The Edison Institute, and one of the first projects undertaken by Ford was to reconstruct the Menlo Park Laboratory complex in Greenfield Village. As part of the dedication ceremonies for The Edison Institute on October 21, 1929, Ford asked Edison to re-create in the reconstructed Menlo Park Laboratory the final moments of the experiment that had led to his development of a successful incandescent light bulb.

Figure 1. On October 21, 1929, Henry Ford and Thomas A. Edison's co-worker Francis Jehl watched while the inventor re-created the famous light bulb experiment as part of the dedication of The Edison Institute in Dearborn, Michigan.

Ford's museum complex was to be different from other museums: it would be driven by its educational mission rather than the intrinsic value of its collections; it would collect the objects of everyday life on an unprecedented scale; and it would focus on the process of innovation and change as a continuing value of American life.

The heroes of Ford's new museum would be such people as Edison, Orville and Wilbur Wright, George Washington Carver, Luther Burbank, and Charles P. Steinmetz. Ford worked long and hard to locate structures related to their lives and moved them to his new museum to take their place as icons of the American spirit of innovation. Edison's laboratory, the cycle shop where the Wright brothers developed the first airplane, Burbank's garden office, the Carver memorial, and Steinmetz's cabin retreat are all part of an effort to memorialize the power of new ideas and new ways of doing things. For Ford this message had a very pragmatic dimension. His goal was to create a museum that would not only record the past but would shape the future as well. It would use the past to encourage visitors, especially the young, to aspire to great achievements of their own. While Ford's goal for The Edison Institute may seem to have been extremely optimistic, it embodies a faith that remains compelling in a society that appears to have lost confidence in its future.

In this year of 1989, the 60th anniversary of the museum, it is extremely appropriate to rededicate ourselves to the founding vision of the museum. Today, our historical perspective is more complex than Ford's, and our way of organizing the delivery of our educational message is quite different. Yet, we at Henry Ford Museum & Greenfield Village maintain a strong faith in the power of this institution to act as a major educational force in the life of our nation. This book is one expression of that belief.

Each of the essays in the book addresses a different aspect of the process of invention as it was carried out at Edison's Menlo Park Laboratory. They are tied together with thoughtful and imaginative precision by Curator William S. Pretzer.

If this book and the reconstructed Menlo Park complex in Greenfield Village can provide enabling energy to new generations by conveying the creative power of Edison and his co-workers, we will have been true to our mission and to the historical traditions of our museum as well.

Harold K. Skramstad, Jr., President
Henry Ford Museum & Greenfield Village

Acknowledgments

A project like this book, especially a collaborative publication produced by a large institution, requires many talents. Many people deserve to be acknowledged for their effort, talent, and enthusiasm.

Inasmuch as this book was inspired by the reinstallation of the Greenfield Village Menlo Park Laboratory, it seems appropriate to begin with those staff members who contributed to that major project. John Bowditch was project manager and chief curator of the 1985–87 reinstallation. His vast knowledge of Edison and technology is evident in the precision and evocative atmosphere of the laboratory complex. His reading of the editor's introduction improved that product, also. Henry Prebys was project manager of the reinstallation of the Sarah Jordan boardinghouse, and Nancy Bryk contributed her enormous talent for interpreting historic interiors as chief curator of that project. Conservator James Burnham coordinated the conservation of hundreds of important artifacts, while Conservator of Historic Buildings Blake Hayes oversaw the physical restoration of both the boardinghouse and the Menlo Park buildings.

Beyond contributing essays that appear in this book, each of the authors, especially W. Bernard Carlson, Paul Israel, and Edward Jay Pershey, offered suggestions, comments, and encouragement that materially aided both the reinstallation project and this publication.

John Tobin researched the historic photographs and coordinated the production of old and new photography with enthusiasm and good humor. Cynthia Read-Miller, Jeanine Head, and Jennifer Heymoss were of great assistance. Photographs of the individual artifacts and of the installation in Greenfield Village were produced with care by Rudy T. Ruzicska, Alan R. Harvey, and Tim Hunter, assisted by Pam Ramirez. C. William Austin of the General Electric Research and Development Center, Joe Cravotta and Eric Olson of the Edison National Historic Site, and Bunny White of the AT&T Archives facilitated the acquisition of photographs from their respective organizations.

Sandy Raidl mastered the production and manipulation of electronic copy for typesetting with greater speed than anyone had a right to expect, greatly facilitating the preparation of this book.

Once again, Sharon Blagdon-Smart has designed an attractive publication that subtly expresses its intellectual underpinnings through its physical structure and visual appeal. Fannia Weingartner is a publications consultant extraordinaire: strategic planner, conceptual editor, proofreader, and production manager. This invention is as much hers as anyone's. Jan Kozora Less and Sarah M. Underhill also proofread.

A number of other people, family members as well as professional colleagues, have patiently tolerated the editor's preoccupation with this project. To all of these people, named and unnamed, I offer my sincere thanks.

William S. Pretzer

Chronology of Thomas Alva Edison's Life

1847 Thomas Alva Edison born February 11, in Milan, Ohio.

1854 Moves with his family to Port Huron, Michigan.

1855 Attends school for three months.

1859 Begins work selling newspapers and candy on the Grand Trunk Railway between Port Huron and Detroit.

1863 Begins employment as a telegraph operator wandering throughout the South and the Midwest.

1867 Finishes his years as a traveling telegraph operator and returns to Port Huron.

1868 Moves to Boston and patents his first invention, an electric vote recorder.

1869 Announces his intention to become a full-time inventor, moves to New York City, and forms first business to manufacture telegraph equipment.

1871 Establishes his own factory and laboratory in Newark, New Jersey, and marries Mary Stilwell.

1874 Develops a quadruplex system for the telegraph.

1876 Establishes laboratory at Menlo Park, New Jersey.

1877 Invents the phonograph and a carbon transmitter for use in telephone.

1879   Develops the first successful electric incandescent lighting system.

1882   Begins work on Pearl Street Central Power Station in New York City.

1884   Edison's wife, Mary, dies.

1886   Edison marries Mina Miller.

1887   Establishes laboratory in West Orange, New Jersey.

1888   Develops his own improved phonograph for commercial production.

1889   Forms Edison General Electric and invents kinetograph, an early motion picture camera.

1892   Sells his interest in electric companies with the formation of General Electric by the merger of Edison General Electric and Thomson-Houston.

1894   Introduces kinetoscope, a motion-picture viewer; the first commercial kinetoscope parlor opens in New York City.

1899   Fails in his efforts to develop a commercial process of magnetic ore separation.

1909   Perfects a nickel-iron-alkaline
storage battery.

1911   Organizes Thomas A. Edison, Inc.

1929   On the 50th anniversary of the
invention, Edison re-enacts his demonstration of the light bulb at the
reconstructed Menlo Park Laboratory
for the dedication of The
Edison Institute.

1931   Edison dies October 18, aged 84.

Introduction: The Meanings of the Two Menlo Parks

William S. Pretzer

THIS BOOK IS ABOUT the daily life and work of the men who took part in one of the most impressive experiences of inventing known in American history—the activities of the Menlo Park Laboratory established by Thomas Alva Edison in New Jersey in 1876. The decision to develop this book grew out of the 1987 reinterpretation of the laboratory reconstructed in Greenfield Village in 1929 on Henry Ford's initiative. The editor's charge to the contributing authors was to explore "the operation of the Menlo Park Laboratory complex, the role of the Menlo Park experience in establishing a model for scientific/technological research, and the social context of living and working in Menlo Park . . . focus[ing] on issues of control and communications in this revolutionary community." This collection of essays aims to bring insights on those issues to an informed and interested lay audience.

Figure 2. *The rural character of the New Jersey countryside surrounding Menlo Park is evident here. The Menlo Park Laboratory is at the far upper left.*

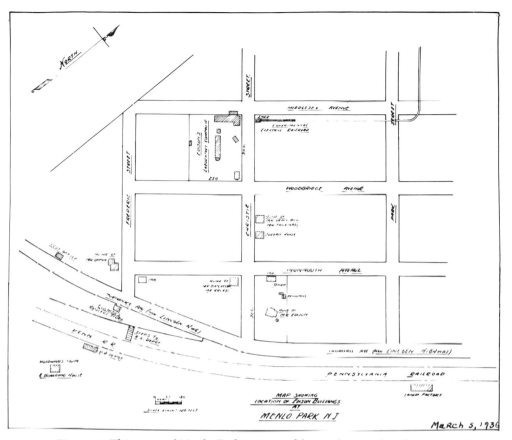

Figure 3. *This map of Menlo Park, prepared long after its abandonment, shows the relationship of the laboratory to the homes of some key workers.*

The laboratory created by Edison in Menlo Park, New Jersey, was one of the wonders of the 19th century. But the lessons we can draw from it in the late 20th century are potentially almost as important as the impact it made on the late 19th century. The experience of successful inventors is especially relevant as Americans look toward the 21st century in search of a renewed sense of technical ingenuity and productive power. American confidence in both have been severely shaken in recent decades. What Edison wrought at Menlo Park may teach us something new today. Certainly many of our most inventive minds believe that Edison provides a model for technological creativity and ingenuity. His name comes up often in a recent collection of interviews with modern inventors. Many of them explicitly view Edison's laboratory as a model of the setting most congenial to the invention process.

This seems surprising, because while so much that is taken for granted today had its beginnings in Menlo Park, little is known about what actually went on there. While the significance and excitement of Edison's achievements there—the phonograph, the telephone transmitter, the incandescent lighting system—readily come to mind from school books and Hollywood movies, the popular conception of Edison and his work is maddeningly vague and even misleading. Memory of historical data can be difficult for those who do not often call upon their historical knowledge. Edison ranked 15th in a recent survey of high school students asked to name influential Americans from the period between the Revolution and the Civil War, even though his work came after that era. Or witness a recent television commercial for beer in which an ersatz Edison explains his light bulb to an interested onlooker, who then says, in effect, "Yes, but Tom we wanted a light beer, not a light bulb." Unfortunately, the bulb this bogus Edison describes is a gaseous bulb, not an incandescent bulb: "The electric current flows through the wire, exciting the gases and creating light."

Much of the popular image of Edison has come from the popular media. Mickey Rooney and Spencer Tracy embody the essence of Edison the energetic, inquisitive boy and the enigmatic, ambitious man. The perception of technological progress as the simple result of the activities of heroic individuals assumes that inventions and technological change are brought about by inquisitive boys and ambitious men. This view ignores vast issues of motivation, inspiration, methods, support, and development for innovative activity. To attribute inventions merely to personal traits and idiosyncratic characteristics overlooks the psychological, economic, technical, and social

Figure 4. *Spencer Tracy as Edison astounds and confuses spectators with a demonstration of the phonograph in the 1940 film,* Edison, The Man.

bases for creativity and technological change. Do inventors share certain intellectual or psychological similarities that should be cultivated? Are there material things one can do to enhance the likelihood of success for a creative process? The concern of this book is to examine the actual social conditions of the Menlo Park Laboratory and to relate them to the key processes of invention. In this way we come to see more explicitly what actually contributed to the productivity of Edison at Menlo Park.

By looking at the Menlo Park experience from the perspectives of psychology, history, and sociology, the essays in this volume explore key aspects of the creative process in a particular place at a particular time. It is our hope that this historical exploration, couched in accessible language, will appeal to a broad public. Many readers may find in this book a different view of Edison from the one commonly held. In the context of actual working days and sleepless nights, the Thomas Edison who emerges here is less a superhuman genius and more an actual person with extraordinary talents. But his talents are definable: an ability to visualize concepts; to think in analogous terms about quite different processes; to organize men and work; to inspire optimism, loyalty, and dedication.

It is really no surprise that Thomas Edison is so well known at a superficial level of American popular culture. He embodies much of what Americans have felt was positive about the national experience. Certainly, as we will see, Henry Ford felt that Edison's value system and accomplishments offered just the model he wanted to foster. Moreover, Edison achieved fame at an early age and lived a long life, much of it basking in the light of his previous achievements. He was by no means ignorant of the importance of personal charisma and cultivated a persona that served his purposes.

Born in Milan, Ohio, in 1847, Edison grew up there and, after the age of six, in Port Huron, Michigan, a typical small-town boy from the Midwest, full of curiosity and mischief. He was not a quick student and his mother took to tutoring him at home. Much, probably too much, has been made of his problems in school. Few boys attended more than a few months of school each year in the mid-19th century, and many went to work as early as Edison did when, aged 12, he took a job as a railroad concessionaire.

It was the beginning of the Civil War and modern means of transportation, communication, and warfare were much on people's minds. Edison had the advantages of living in Michigan, far from the battlefront, and of being too young to enlist. His association with railroads led him to a familiarity with the telegraph, then the most advanced form of communication. Since telegraphy was based on electricity, a relatively new frontier of scientific and technical research, Edison was ideally situated to make something of himself. It was an expansive era with plenty of opportunities open to men of ambition. Edison spent the Civil War years as a tramping telegrapher, moving from office to office, following his own whim and the desires of his employers. In each place he took advantage of his access to telegraph technicians and materiel to expand his knowledge and experience.

Two essential traits surfaced early in his life: an optimistic attitude and a restless urge to make a difference. If there was a characteristic that set Edison apart from his fellows, it was his conviction that he could devise better ways to do things than were customary. After the war, Edison attempted to improve his status from that of a mere operative to that of an inventor within the telegraphic industry. In 1868 he arranged to have a Boston businessman finance some experiments and a year later he announced that he would devote himself full time to inventing and received his first telegraph patent.

In that same year, 1869, Edison was named superintendent of a New York City telegraph company, established to send out continuous business reports—including the price of gold—and later was in a position to go into partnerships inventing and manufacturing telegraphic apparatus. His success at developing new apparatus, at attracting able men to work with him, and at making connections with influential business leaders allowed him to achieve the establishment of a full-time research and development lab, what he termed an "invention factory."

From 1876 until 1882, Edison's efforts were concentrated in a small group of buildings in Menlo Park, New Jersey. The results were nothing less than earth-shattering: in six short years, Edison obtained more than 400 patents, some of which represented the most influential technologies of modern America. He continued to work out various telegraphic problems; he invented the phonograph; he perfected a carbon telephone transmitter; and he was responsible for perhaps the most astonishing development of all—the practical incandescent light bulb and electrical generating and transmitting system. All his work focused on projects that would result in actual changes in everyday life in homes or offices.

Figure 5. *Edison proudly poses beside his favorite invention, the phonograph, in 1878.*

From 1882 until 1886, Edison concentrated on putting his electric lighting system into commercial use. The first urban application came with his Pearl Street Station and the electrification of two square miles of downtown Manhattan. In 1887 Edison opened a new research and manufacturing facility in West Orange, New Jersey, several miles north of Menlo Park. With his base in West Orange, Edison continued his inventive activity. He improved his own phonograph into a commercially viable product, both as a business dictation machine and as a phonograph. From 1889 until 1929 his company was one of the major producers of commercial home phonographs and musical recordings. He created the kinetograph and kinetoscope (an early motion picture camera and viewer), dramatically fostering the development of the motion picture industry. While he failed to develop a commercially successful process for magnetic ore separation, he did succeed in inventing the nickel-iron-alkaline storage battery in 1910. During World War I he served as chairman of the Naval Consulting Board, a largely honorific organization that drew public attention to the applications of scientific research to wartime uses. Although he continued to receive patents on inventions until his death in 1931, Edison's last years were spent more in participating in family and civic rituals and basking in public adulation than in carrying on productive research. He died, aged 84, the possessor of 1,093 patents, probably the most famous American inventor since Benjamin Franklin.

At Menlo Park, Edison received funding from the Western Union Telegraph Company (one of the great American monopolies) to erect a laboratory building and stock it with the equipment and supplies necessary for research in electrical and chemical engineering. He enlisted several men from his Newark shop to accompany him and added to his work force as need dictated. Initially, the key figures in the Menlo Park Laboratory included Charles Batchelor, a talented collaborator and adept experimenter; John Kruesi, a skilled machinist and foreman of the machine shop; and William Carman, who served as secretary and accountant. Later, Edison added individuals with specific areas of expertise. Batchelor was joined in the lab work by Francis Upton, a university-trained mathematician; by John Lawson, an assayist; by Alfred Haid, chemist; and by Francis Jehl, Martin Force, and other laboratory assistants. Skilled craftsmen such as Milo Andrus the carpenter; Ludwig Boehm and William Holzer, glass workers; and machinists like Charles Dean were also brought on. Finally, Edison's business interests were served by additional people like Stockton Griffin, his personal secretary, and Ernest Berggren, another bookkeeper. Thus, Edison enlarged his staff along lines already determined rather than changing its character. By adding new blood during the Menlo Park years he created both a critical mass of energetic, ambitious minds and a fresh mixture of personalities.

Simultaneously Edison expanded the physical components of the complex. The first building, a two-story frame structure, originally combined office, laboratory, and machine shop. He added a glass house and photographer's studio, a carpenter's shop, a carbon production shed, and a blacksmith's shop, all of wooden frame construction. In 1878 Edison began the construction of a machine shop and a combined business office and library, both made of brick. Eventually the compound included, as well, a shed which stored an experimental electric train locomotive. At its most active, the Menlo Park compound was a compact group of buildings surrounded

Figure 6. *From the front of the laboratory, one could see past the corner of the brick office and library building, left, to Sarah Jordan's boardinghouse and down the hill to the windmill and the Edison home at the far right.*

by a white picket fence located on the top of a knoll overlooking the small village of Menlo Park and the surrounding countryside.

The village of Menlo Park had been chosen for a variety of reasons. It offered access to resources and labor, yet lacked the intrusive and distracting traits of a city. Land owned by William Carman's family was available and favorably priced because of the economic recessions of the mid-1870s. Situated some 25 miles south of Manhattan on the Pennsylvania Railroad line connecting New York City to Phildelphia, Menlo Park was no more than a cluster of dispersed houses served by a railroad station and a combined tavern and grocery. Even in 1880, when Edison's laboratory was at the height of its operation, the total population of the village was only 200. But of that number, about 75 were directly associated with the laboratory and others represented their families and servants. Edison was able to concentrate his work force and have direct connections with most of his employees. Master and workers lived close to one another.

Indeed, when Edison created the laboratory he created a small company village. At the bottom of the ridge ran the Lincoln Highway and the Pennsylvania Railroad tracks. Immediately adjacent were the tavern and several houses. Halfway up the hill was the largest house (and the only one with a windmill to pump water) which Edison purchased for his family. Further up the hill were more houses, and at the top, the small mystical research enclave. One might even, with some hyperbole, go on to attribute a religious character to this small community. Here Edison was the sage and the workers

Figure 7. *The main laboratory building at Menlo Park hardly looked like the center of an important research laboratory when this photograph was taken in 1878.*

Figure 8. *Edison poses with co-workers and two of his children, who were affectionately known as Dot and Dash.*

Figure 9. *Sarah Jordan's boarding-house was home to as many as 16 men at one time. (Inset) Mrs. Jordan was a widow making a living in one of the few respectable ways available to unmarried women of that era.*

Figure 10. *The Edison home in Menlo Park was a substantial dwelling maintained by several black servants.*

lived, vassal-like, in close proximity. Their positions were entirely dependent upon Edison. Neither the Edisons nor the workers had much contact with the local farmers: the peasants, to extend the metaphor, were not part of this religious community. The religion here was empiricism and science, and the laboratory building, constructed at the top of the hill with a clerestory window on the gabled front, was the cathedral.

While the laboratory complex itself was a male domain, this was not a monastic community. Several of the men, Edison and Upton among them, brought wives to live in Menlo Park. Several of the householders employed male and female servants, including Irish immigrants and blacks. Some of the workers roomed with local families or commuted on foot from the surrounding area.

Edison was even instrumental in establishing a boardinghouse for his employees in the community. Sarah Jordan was a 41-year-old widow with a 13-year-old adopted daughter. A distant relative of Edison's wife, she had been the family's neighbor in Newark. In 1878 Edison enlisted her to move to Menlo Park to operate a boardinghouse for his employees.

The work force was just beginning to grow and many of the single men moved into Sarah Jordan's boardinghouse, located down Christie Street some 50 yards from the entrance to the laboratory complex. From 10 to 16 men roomed and ate meals provided by Sarah, her daughter Ida, and the maid, Kate Williams. The women occupied one half of the downstairs and tended to domestic chores. They cleaned the house, did laundry, and cooked for hearty appetites. The raucous nature of the men's jokes and horseplay was only slightly tempered by Sarah Jordan's stern but motherly demeanor and reportedly delectable pies.

From the front parlor or their rooms, the men could clearly see (and be seen from) the laboratory grounds and the wooden boardwalk leading from the railway station past Edison's house up the hill to Sarah Jordan's and on to the laboratory itself. Not that they spent much time in the boardinghouse; but its convenient location meant that even there they were not far from work nor from Edison's attention. Of course, there were literally only two other places to go without leaving the community for a several-mile walk to the town of Metuchen. Local recreation was centered around the combined railroad station, telegraph office, and post office building, and the tavern by the railroad tracks known as the Lighthouse, which was operated by a garrulous Scotsman named Davis. Not only did this tavern provide growlers of beer for the workers as well as pails of food for the late-night sessions in the laboratory, but it also sported the only billiard table in the area.

Menlo Park offered equally few distractions for Edison himself. In 1871 he had married Mary Stilwell, a Newark teenager he had employed, and by the time they moved to Menlo Park they had two children. A third was born there in 1878. Besides her children and a sister who came to live with her, Mary Edison found little of interest in Menlo Park. Nor did her husband find much time for her. It was thus an unhappy Mary Edison who took ill and died in 1884, leaving Edison's mother-in-law to care for the children in her New York City home.

Edison had been young and unknown at the time of his first marriage in 1871. By 1884 he was rich and famous. Attracting women was no problem; finding time and inclination to court them was another matter. Through a

Figure 11. *(Left) Edison in 1879.*

Figure 12. *(Above) Edison in 1882, looking every bit the successful entrepreneur.*

series of connections, Edison was introduced to Mina Miller, the daughter of a prominent Akron, Ohio, tool manufacturer and philanthropist. He married her in 1886. They had three children between 1888 and 1898. Edison and his second wife lived in an enormous mansion, called Glenmont, near his West Orange laboratory. While Mary, who had come from a lower-middle-class family, had been unassertive, Mina was worldly and opinionated. For the rest of his life, Edison could rely on Mina to have strong views and to serve as a firm advocate on his behalf.

Edison had established Menlo Park in 1876 while participating as a minor inventor in the Philadelphia Centennial Exposition. In 1887 he established his much larger research and manufacturing complex in West Orange, shortly after marrying his second wife. His first research laboratory was in a rural area; his second on the fringes of the growing industrial complex surrounding New York City. His first wife was naive; his second sophisticated. Edison had moved in both professional and personal ways from the country-boy-turned-inventor into the rough-hewn Gilded Age entrepreneur. Menlo Park had been the major step in his own transformation. Its social and technical characteristics had reflected and influenced his life in a variety of ways. Edison moved into Menlo Park with unabashed enthusiasm; he left with the world at his feet.

Presenting the Mystique of Edison's Inventiveness

This book is very much like an invention. It first existed as a problem, or set of questions, and some notions of how to resolve them. It has been developed into a physical form composed of ideas that are supposed to accomplish specific goals in a particular way. In the case of this publication, the initial challenge was to describe the Menlo Park experience in light of new interpretations of the history of 19th-century science and technology. The importance of the reconstruction of the Menlo Park complex in Greenfield Village is widely recognized by scholars and the general public. It is one of the few research laboratories to be re-created as part of a historical presentation. It is a widely recognized symbol of American ingenuity. This publication is, in part, the product of a 90-year history that links the names and accomplishments of two famous innovators, Thomas A. Edison and Henry Ford.

Henry Ford worked early in his career as an engineer for the Edison Illuminating Company of Detroit, managing steam-driven generators of electricity. He and Edison were introduced to one another casually in 1899 and Ford later credited Edison with encouraging his experiments with a gasoline-powered internal combustion engine automobile. Some years later, after Ford had made a reputation and a fortune by manufacturing cars, he and Edison became good friends, sharing a great deal of public acclaim for their technological accomplishments. When Ford founded the indoor-outdoor educational complex he envisioned as a new type of socially relevant learning institution in 1929, he named it The Edison Institute. This was his way of honoring the man whose accomplishments and values he held in the highest esteem.

At the heart of the outdoor museum, named Greenfield Village, stood a reconstruction of Edison's Menlo Park Laboratory complex. The General Electric Company donated the only original building from the laboratory complex—the glass blowing shed—that it had moved to a company-sponsored park in Parsippany, New Jersey. Based upon a study of the remaining structural ruins in New Jersey, some building elements scavenged from the surrounding countryside, photographic evidence, and oral reminiscences of Edison and his co-workers, Ford constructed six buildings to represent the original site. The one Edison-related building still left standing in Menlo Park, the Sarah Jordan boardinghouse, was moved *in toto* to Ford's historic area.

At the core of Ford's attempts to document American ingenuity lay a huge collection of objects once used or produced by Edison, his associates, or companies bearing his name. These artifacts included a variety of scientific apparatus, patent models, and commercially manufactured Edison products dating from the Menlo Park era. Thus, physically as well as intellectually, the institution was initially based upon a reverence for Edison's technological and social philosophy, his material contributions to science and invention, and his influence upon modern life through the development of sound and light systems.

The Menlo Park reconstruction in Greenfield Village has been a favorite of visitors for decades. Upwards of a million people a year tour the complex. Most come with a common textbook image of Edison as a slightly eccentric, highly motivated technical genius. Many think of his inventions as the direct result of his personal knowledge, imagination, talent, and perseverence. For decades Menlo Park itself has been generally referred to as an invention fac-

Figures 13 and 14. *Well before Henry Ford decided to reconstruct the Edison laboratory only ruins remained in Menlo Park, New Jersey.*

tory, but the actual meaning—beyond the simplistic one of a place for the mass production of patented inventions—has never been clear.

Recently, public presentations at the site have concentrated on communicating the historical impact of the idea of an invention factory. Edison's major contribution to modern life came in his effective combination of science, technology, and business. His ability to orchestrate ideas, men, materiel, and financing set him apart from his contemporaries. It was through his ability to subdivide the invention process and devote attention to the systematic application of research and development methods, rather than through the development of any single invention or even group of technologies, that

Figure 15. *To mark the Golden Jubilee of Light in October 1929, Henry Ford, United States President Herbert C. Hoover, Edison, and Francis Jehl (right to left) gathered in the reconstructed Menlo Park Laboratory in Greenfield Village for a re-creation of the light bulb experiment.*

Edison revamped the modern world in a lasting way. Rather than trying to focus public attention on the achievements of technological virtuosity displayed by Edison's team, the interpretation made in Greenfield Village points toward Edison as one of the architects of 20th-century capitalism, a socio-economic system that retains power in part through the definition and coordination of science and technology.

If Menlo Park, New Jersey, created great inventions and a model for later researchers, then Menlo Park, Greenfield Village, conjures up images of technical curiosity and ingenuity, teamwork and individual brilliance, business and experimentation. Both have been adept at creating their own myths for popular consumption. More than a century apart, they both have proven to be of immense public interest; both continue to inspire notions of great social benefits from science and technology.

Figure 16. *The reinstalled office at Menlo Park, with its accountant's desk, draftsman's table, and stacks of correspondence, reflects Edison's need to control his finances, his patents, and his public image.*

Figure 17. *The second floor of the main laboratory building suggests a key element of Edison's leadership: his ability to gather together skilled workers, supplies, and equipment.*

Figure 18. Sarah Jordan's boardinghouse in Greenfield Village interprets the domestic work performed by women in support of the inventive work being carried out in the laboratory up the street.

Figure 19. The rooms at Sarah Jordan's were clean but sparse, befitting the transient lifestyle of many of the men who lived there.

New Impressions of Menlo Park

The essays in this volume describe the actual work performed at Menlo Park; evaluate the important elements of Edison's refinement of invention processes; and assess the relationship of the Menlo Park experience to the larger history of research and development. For all of the conclusions and lessons supposedly drawn from the Menlo Park years by scholars and laymen alike, it is surprising that so little has actually been published about the realities of daily life there. This volume begins to redress that omission.

Bernard S. Finn's essay opens the discussion by identifying in some detail the types of men who worked with Edison and the kinds of work they performed. First, Finn documents the range of specialists required by Edison's operation and the varied ways they came to be in his employ. Second, he makes clear just how eager young men were to take advantage of the opportunity offered by Edison's advanced research work, and how much they were willing to sacrifice to work on the cutting edge of a burgeoning technology. Finally, Finn suggests that Edison's employment patterns reflected a calculated effort to influence his work force through creating bonds of personal loyalty rather than setting up bureaucratic procedures and controls.

Finn's essay, while contributing significant details on its own, also provides a convenient introduction to Andre Millard's analysis of craft culture at Menlo Park. Millard points out that Menlo Park, while representing a new kind of establishment aimed at technological innovation, still was built upon the traditional methods of craft production as exemplified by the performance and work culture of skilled machinists. The inherent contradiction in the term "invention factory" lies at the heart of Millard's analysis. How was it that Edison attempted to reconcile his desire for systematic attention to imaginative and creative thinking within an industrial system in which creativity was spasmodic and factory work routine? Millard argues that Edison relied upon a comfortable and well-known set of practices by which machinists (and other skilled workers) controlled the shop floor operations and kept their superiors at arms' length. As long as Edison could rely on his assistants to maintain orderly work, he had no need to insist upon the discipline demanded by factory owners.

Paul Israel maintains that Edison formed his approach to the organization of innovation from his experiences in the telegraph industry. Telegraphy had all the aspects of electrical, chemical, and mechanical engineering centered in the machine shop and laboratory that Edison went on to apply to broader development questions. In developing methods for attacking problems, Edison relied on history, namely, his own past experiences. Both Millard and Israel see the roots of innovation in Edison's ability to utilize the best of the known and familiar techniques of organization, supervision, and accountability.

W. Bernard Carlson and Michael E. Gorman, on the other hand, find Edison's major strength in his capacity to think in conceptual terms and utilize, perhaps unconsciously, a series of building blocks and models to develop a picture of a mechanism that could perform a specific function, in this case the carbon telephone transmitter. It is the interplay between the conception of how a "thing" operates to achieve its goals and the parts that might go into such a "thing" that constitutes the inventor's great mental capacity. In this version of the mental processes of inventors, Carlson and

Gorman contribute a view of creative thinking that is more substantial (and perhaps more teachable) than the more popular notion of inspiration and perseverance, yet more abstract (and therefore applicable) than a detailed description of one invention in which the concepts are not transferrable to another project. Carlson and Gorman make extensive use of the telephone artifacts to trace one of Edison's more important inventive activities.

Communicating about three-dimensional objects, unfamiliar technologies, and abstract concepts involves particular skills. Edison was working in fields where there were few kinds of "shorthand" or commonly shared notions about the theory or reality of electro-chemical-mechanical practices. Edward Jay Pershey's essay on the phonograph demonstrates how Edison, who was known for his ability to sketch ideas, relied on visual communication as a primary way of thinking through problems and relaying ideas to his associates. Drawing became a vernacular language for the Menlo Park community. Pershey also makes extensive use of the extant record by showing in great detail the evolutionary drawings of the phonograph. In this way

Figure 20. In 1930, as today, Ford's detailed reconstruction bore an uncanny resemblance to the original site.

he contributes to our understanding of the steps in invention, the actual process of developing the phonograph, and the importance of imagery in conveying technical information. It may well be that Pershey's attention to the drawings themselves suggests a method of illustrating the building-block model provided by Carlson and Gorman.

Finally, David A. Hounshell offers an assessment of the importance of the Menlo Park experience to the future of American research and development. Hounshell finds in the exigencies of daily life and personal motivations clues to Edison's ability to move between scientific and technological/industrial cultures. Certainly, it is this ability that has characterized much of 20th-century corporate-controlled research and development. (It would be interesting to speculate on the subsequent influence of government-sponsored research and the role of universities in linking government and corporate research programs.) Hounshell thus provides a means of appreciating how Edison's small laboratory, based on traditional work practices but with a view to the future, could contribute so much in the way of models for future research communities.

Edison's penchant for thinking through technological questions in the form of technical systems matched his ability to establish astute professional connections. Both his technical systems and his personal networks mirrored the requirements and opportunities of big business. A century ago Edison pioneered the establishment of the symbiotic economic and institutional relationships between business and technological change that would later become commonplace. He showed us the way to the 20th century, for good and ill.

As we look at these individual essays, each a contribution to the study of Edison and of invention processes, we would do well to consider how these processes compare to our contemporary versions of research and development. Have we developed the proper teams or conceptions of teamwork? Have we developed new methods to engender the loyalty, creativity, and education that the artisan tradition once did? Are we able to connect abstract concepts with particular visual images, or have our contemporary, so-called visually literate media actually debased the communicative nature of imagery? Do we, in fact, provide the necessary motivations and inducements for individual and social commitment to technological progress? The Menlo Park experience of Thomas A. Edison raises many questions pertinent to today's society, and the essays in this book have significant implications for contemporary innovators.

Bibliography

More than 75 books and dozens of articles have been written specifically about Thomas A. Edison. The basic facts of his life can be found in these biographies: Matthew Josephson, *Edison: A Biography* (New York: McGraw Hill, 1959) and Robert Conot, *A Streak of Luck: The Life and Legend of Thomas Alva Edison* (New York: Seaview Books, 1979). Provocative interpretations of the meaning of Edison's career are presented in David Nye, *The Invented Self: An Anti-Biography, from Documents of Thomas A. Edison* (Odense, Norway: Odense University Press, 1983) and Wyn Wachhorst, *Thomas Alva Edison: An American Myth* (Cambridge, Mass.: MIT Press, 1980). Most directly relevant to an understanding of Menlo Park are Francis Jehl, *Menlo Park Reminiscences*, 3 vols. (Dearborn, Mich.: The Edison Institute, 1934–41) and Robert Friedel and Paul Israel, with Bernard S. Finn, *Edison's Electric Light: Biography of an Invention* (New Brunswick, N.J.: Rutgers University Press, 1986). The history of Henry Ford's Edison Institute is presented in Geoffrey C. Upward, *A Home for Our Heritage: The Building and Growth of Greenfield Village and Henry Ford Museum, 1929–1979* (Dearborn, Mich.: The Edison Institute, 1979). Essential for understanding the context and continuing importance of history museums like Henry Ford Museum & Greenfield Village is Michael Wallace, "Visiting the Past: History Museums in the United States," in Susan Porter Benson, Stephen Brier, and Roy Rosenzweig, eds., *Presenting the Past: Essays on History and the Public* (Philadelphia: Temple University Press, 1986): 137–164.

Working at Menlo Park

Bernard S. Finn

IN THE COURSE OF the approximately five years that his Menlo Park Laboratory functioned as a center for invention, Thomas A. Edison hired more than two hundred employees. Some of these stayed for only a few weeks; others remained with him or with Edison companies for the rest of their careers. All of them contributed, in one way or another, to the mystique of the laboratory and to the prodigious output associated with its leading figure.

Who were these people? What brought them to Menlo Park? Why did they stay? What did they do? And why did they leave? With the help of reminiscences and biographical accounts, as well as such documentary evidence as time sheets and payroll lists, this essay attempts to answer these questions and provide a view of the laboratory from the vantage point of the

Figure 21. *This painting of Menlo Park in the winter of 1880–81 shows both the arrangement of the buildings in the complex and their rural surroundings.*

Figure 22. *In this 1877 illustration Edison wears a cap and scarf akin to those commonly used by skilled artisans working at Menlo Park.*

workers. The result, one hopes, will contribute to a better understanding of the laboratory's unique character.

The picture presented is to some extent impressionistic, given the available evidence. Good biographical material exists for only a few employees, though usually for the more important ones. Time sheets and payrolls provide spotty coverage, but often yield substantial information about specific jobs being worked on. An unexpectedly profitable source was the evidence given at a series of patent interference hearings held in the course of suits brought by or against Edison. Some of the Menlo Park employees were called as witnesses and were asked detailed questions about their life and work. From these and other sources, direct knowledge of geographic origins, prior training, motivations for coming to and leaving Menlo Park,

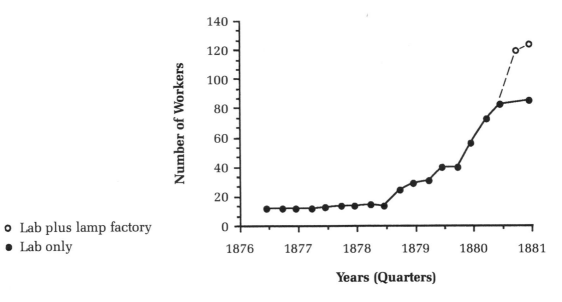

o Lab plus lamp factory
• Lab only

Figure 23. This graph demonstrates the growth of the Menlo Park work force between 1876 and 1881.

Figure 24. Taken on the front steps of the main laboratory building around 1880, this photograph shows Edison with several of his laboratory associates and his father. Top row from left: Albert Herrick, Francis Jehl, the elder Edison, George Crosby, George Carman, Charles Mott, John Lawson, George Hill, and Ludwig Boehm; middle row: Charles Batchelor, Edison, Charles Hughes, and William Carman; bottom row: William Holzer and James Hipple.

and the kinds of lives led there is available for a modest percentage of those who worked with Edison. In all, there is enough evidence to be persuasive, and enough information missing to leave room for further questions.

When Edison decided to establish his own laboratory he had a very special notion about what it would be. He had limited resources. And he had some particular projects in mind. These conditions affected the numbers and types of workers he would hire. Because the laboratory was to be a direct extension of his own inventive genius, he wanted people who would execute his ideas rather than originate their own. At first this meant people who were good with tools or good at following directions, but as the number and diver sity of projects increased, and as the resources to pursue them were made available, he added workers with other skills and began to give some of them limited autonomy.

A graph of employment at Menlo Park reflects this evolution. A force of a dozen or so remained relatively constant during 1876–77, the period when the laboratory was working on the telegraph, phonograph, electric pen, and early telephone projects. There was a noticeable jump in the fall of 1878, when Edison began serious work on the incandescent lamp, and a gradual increase through 1879 as this work branched out and activity on the telephone became more pronounced (Fig. 23).

The number of workers continued to grow in 1880 as Edison prepared for his second and more elaborate demonstration of the lighting system on the Menlo Park grounds, and as he prepared for the lamp factory that would open in October. In mid-November the payroll of the lamp factory was officially separated from that of the laboratory, even though there is evidence that some of those listed with the laboratory were spending most, if not all, of their time at the factory, or on factory work at the laboratory. (For instance, lamps were tested there until March 1881.) Edison left Menlo Park in April 1881, marking the end of the period covered by this essay, though others remained active there for a few months and the factory stayed in operation for another year.

In line with his particular needs, Edison's original force consisted of a couple of machinists, some general assistants, and a draftsman. As is well known, when he began his lamp researches he looked for someone who could do a literature search and hired college-trained Francis Upton. With increasing need to analyze materials, he engaged German-trained chemist Alfred Haid early in 1879. Later, responding to a demand for glass tubing for the Sprengel pump and for the light bulb itself, he hired a glass blower, first on a part-time and then on a full-time basis.

Let us now look more closely at these people. Who were they? Where did they come from? The original group came with Edison from Newark. Six can be identified specifically: James Adams, Charles Batchelor, William Carman, John Kruesi, Charles Stilwell, and Charles Wurth. Charles Edison joined his uncle a few months later. James Bradley, Charles Dean, and John Ott, who had all been with Edison at Ward Street in Newark, rejoined the group in the new facility—Bradley probably in 1877, Dean and Ott in 1878. With the exception of Edison's nephew Charles, all had been recruited locally. Adams, Batchelor, Kruesi, and Wurth previously had been employed near Newark, having originally come from Europe; the others had grown up in the immediate area. Later additions to the work force followed the same pattern, though the amount of information available about these men is much more limited. About 70 additional employees were hired through 1879. Among the 20 who have identifiable origins, 6 came from Europe and 9 from New Jersey.

Significantly, most of the men with specific skills came from Europe or the British Isles. Kruesi had been apprenticed to a clockmaker in his native Switzerland; Wurth, also Swiss, was a machinist. In 1879 Edison hired Alfred Haid, a German, as a chemist; Alexander Mungle (who had been apprenticed to a firm of engine builders in Scotland) and Jim Holloway (who had served a seven-year apprenticeship to an engineering firm in London), as machinists; and Ludwig Boehm (who had had extensive training and experience in his native Germany) as a glass blower. William Andrews had received a good education in science and electricity at an academy near Bath, where he had also served as headmaster for a decade before becoming involved in the manufacture of firearms. When he came to Menlo Park in November 1879, it was in the capacity of a skilled machinist.

The exceptions—the college graduates employed at Menlo Park—also fit a pattern. Francis Upton, who arrived in 1878, had a strong background in mathematics, as did Charles Clarke, his classmate at Bowdoin College, who came early in 1880. Physicist Edward Nichols appeared later that same year. Those with nontechnical backgrounds were William Carman, a business college graduate who had worked as a bookkeeper before joining Edison at the Ward Street establishment in Newark, and Samuel Mott, who had spent three years at Lehigh University and two months at Princeton before becoming a patent draftsman for Edison.

About most of the rest we know very little. Martin Force may have had some background as a carpenter when he was hired at the age of 24 in 1875; he went on to become a general assistant on a variety of projects. Francis Jehl, who would later have similar duties, had studied some at Cooper Union, spent a year in the Western Union shops, and been a law clerk when he arrived at Menlo Park at the age of 19. William J. Hammer had graduated from high school, attended some technical lectures at the university level, and worked for a year in Newark (for Edwin Weston) before he started with Edison in December 1879 at the age of 22. None of the others for whom we have information had any technical background, and it seems reasonable to presume that this was true for virtually all in the remainder of the work force. They would receive their training on the job.

Why did they come? Occasionally Edison would advertise for specific needs. In the spring of 1879 he placed notices in the *New York Herald* for an instrument maker and a chemist, and in a German-language paper (and perhaps elsewhere) for a glass blower. (As a result, he hired Haid and Boehm.) But most employees apparently came on their own initiative.

For many who lived in the area, working at Menlo Park was just a job, and not infrequently they applied because a brother had already been hired there. A large proportion of the remaining employees came because they wanted to be trained in electrical technology. It was not unusual for men in this group to start without being paid, sometimes working for several months before being given a salary.

John (later nicknamed "Basic") Lawson can serve as an example. On January 6, 1879, Lawson, then 22 years old, wrote from Vermont, volunteer-

Figure 25. *Francis Upton, Edison's university-trained mathematician.*

Figure 26. *Alfred Haid, chemist.*

Figure 27. *William Carman, accountant.*

Figure 28. *Charles Batchelor, Edison's closest laboratory associate.*

Figure 29. *John Kruesi, master machinist and foreman of the machine shop.*

Figure 30. *Ludwig Boehm, a German-born glass blower who left Menlo Park after a dispute with Edison and his co-workers.*

Figure 31. *John Lawson, a young man who came to work for Edison in hopes of learning to be a chemist.*

Figure 32. *Charles Clarke, a young college-educated friend of Francis Upton who came to work at Menlo Park, eventually became a prominent electrical engineer.*

ing his services: "Nearly four years ago I began to feel the want of an education, and commenced to study, devoting what time I could get in the evening to that purpose." But, he explained, he suffered from lack of laboratory facilities. He went on, "Can you not help me?. . . .I care not what the work is if I can only have a chance to study. . . .I wish to become a chemist. Had I the means I would devote myself to chemistry. . . .I am willing to do anything, dirty work—become anything, almost a slave, only give me a chance to pursue the studies that I love." Edison replied that Lawson could start in the chemical laboratory at $5 per week. He was there by January 16.

On the other hand, a gentleman named F. Hoffbauer, after describing the experience he had gained in Germany, wrote, "I have an invention mind, which in the service of a great genius will be able to perform uncommon." Edison responded, "If you are a good workman at the lathe, can give you a situation as such, but have no position open as an assistant as I have all I desire." Hoffbauer's name does not appear on the Menlo Park rolls.

A month later John Segredor also described himself as an inventor. Either the tone was more pleasing or the closing remark that he would be happy to come for the price of his board struck Edison's fancy, for he joined the staff a month later. Joseph Harbeson and Otto Krebs were apparently of some interest to Edison, for he asked them how much they would like to be paid. Harbeson, with some hesitation, suggested $12 per week, while Krebs made the mistake of asking for $6 in addition to lodging and board. Neither got a job. George Crosby was given a chance at Menlo Park as a favor to a *New York Herald* reporter. He worked for six or seven months at no salary (he did receive money for board) before drifting away. It seems to have been more usual for a young man to be picked up on the payroll *after* he had proven himself. This was true of Albert Herrick, who first came to Menlo Park with his mother, who was writing a magazine article about the complex. He worked more than four months at no salary, then eight more at $5 per week. When he left it was to attend Stevens Institute to get a degree in mechanical engineering. Wilson Howell was so impressed by the lighting displays of December 1879 that he came back the next day to plead for the chance to work without salary and ended up as a long-time employee. Samuel Mott came to learn about the electrical business. He later testified that he would have paid Edison for the opportunity, though in fact when offered $5 per week he was bold enough to talk "the Wizard" up to $7.

Often cited are those who came to work with Edison and later became well known for their scientific or technical achievements. Two qualified for that distinction during the five-year period being studied here. One was Edward Acheson, inventor of carborundum, who left school at the age of 16 and had several technically oriented jobs before joining Edison in September 1880 at the age of 24. He was employed for a little over a year (the last few months with Batchelor in Europe) before leaving after a disagreement over pay. The other was physicist Edward Nichols who, after he had received his Ph.D. from the University of Goettingen and spent a year as a fellow at Johns Hopkins University, was persuaded by Upton to come to Menlo Park to take charge of the testing department. He stayed through the following spring, then left to begin a distinguished teaching career, most of which b at Cornell University.

Thus it can be seen that the publicity surrounding Edison p

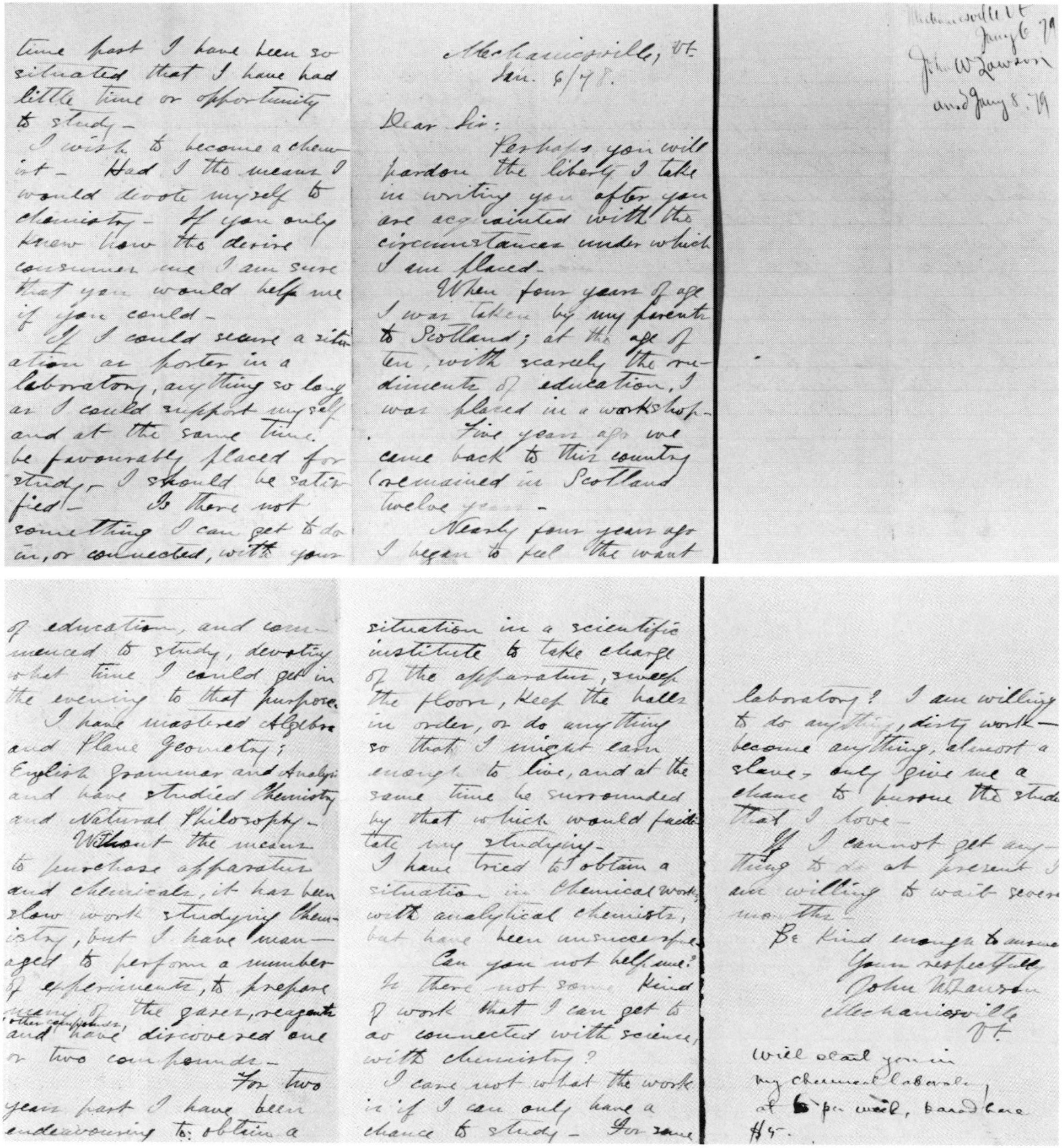

Figure 33. This letter from John Lawson to Edison asking to be taken on as a chemist at Menlo Park indicated that the writer was willing to perform menial tasks for the chance to be near work that interested him [pagination added].

with a ready pool of applicants. At a time when academic training in practical electricity was not easy to find, a significant number of those who sought to come to Menlo Park were anxious to learn about the new technology. This inevitably affected their attitude towards their jobs and the general character of the workplace.

Why did they leave? The answer to this question is much less clear. In later years, Edison was notoriously tolerant of relative incompetence in his assistants. There are a few known instances of someone being fired for stealing or some other overt act, but it is reasonable to suppose that most of those who left did so of their own accord. Among those who were hired in the period 1876–79, approximately half had gone before the end of 1880. Some, like Acheson and Nichols, got the training they needed and probably had never planned to stay any longer. But as far as the others were concerned, with an average tenure of under three months there was hardly enough time for any real training, and we must assume that they left because they found the atmosphere of the laboratory too unconventional, or the hours too long, or the situation generally uncongenial.

What did they do? A few, like Boehm the glass blower or Haid the chemist, had quite specific jobs. But most could be expected to work on almost anything, and often several things during the course of the week. Time cards for a typical week in August 1878 (ending on the 30th) show Milo Andrus with eight different tasks, most but not all related to his carpentry skills; Charles Batchelor on five projects (megaphone, telephone, voltameter, phonograph, aurophone); and the rest individually engaged on anywhere from four to ten projects. The only employees confined to a single area were Charles Edison ("photographing") and N. Ruffner (who worked only one day).

The workweek consisted of six ten-hour days. For the week cited above, most of the dozen men on the payroll are marked down for 60 hours or even less. Four worked official overtime, from 6 to 13 hours. In December, with work on the lamp well underway, the numbers increased. Out of 21 employees, only six worked 60 hours or less; the average of the rest was just under 80 (the highest was 94). The number of tasks, however, was down, with most of the men working almost exclusively on the electric light.

In 1879 the personal workload remained much the same. From time sheets in June and October we find that, although the staff size had increased, approximately two-thirds of the men were working an average of 80 hours a week. The trend to specialization had intensified, and most of the men worked either on the telephone (which was absorbing the bulk of the laboratory's energy) or the electric light.

The pattern of work at Menlo Park was most graphically described in the oft-quoted article that appeared in the *New York Herald* in January 1879:

The ordinary rules of industry seem to be reversed at Menlo Park. Edison and his numerous assistants turn night into day and day into night. At six o'clock in the evening the machinists and electricians assemble in the laboratory. Edison is already present, attired in a suit of blue flannel, with hair uncombed and straggling over his eyes, a silk handkerchief around his neck, his hands and face somewhat begrimed and his whole air that of a man with a purpose and indifferent to everything save that purpose. By a quarter past six the quiet laboratory has become transformed into a hive of industry. The hum of machinery drowns all other sounds and each man is at his particular post. . . . Edison himself flits about, first to one bench, then to another, examining here, instructing there; at one place drawing out new fancied designs, at another earnestly watching the progress of some experiment. Sometimes he hastily leaves the busy throng of workmen and for an hour or more is seen by no one. Where he is the general body of assistants do not know or ask, but his few principal men are aware that in a quiet corner upstairs in the old workshop, with a single light to dispel the darkness around, sits the inventor, with pencil and paper, drawing, figuring, pondering. In these moments he is rarely disturbed. If any important question of construction arises on which his advice is necessary the workmen wait. Sometimes they wait for hours in idleness, but at the laboratory such idleness is considered far more profitable than any interference with the inventor while he is in the throes of invention.

Certainly there is much truth in this picture, though evening work was more likely to be just an extension of daytime activity. It was, after all, difficult to work an 80-hour week—usually with Sunday off—without getting in a lot of both daylight and evening hours.

Charles Clarke, who arrived at Menlo Park on Sunday, February 1, 1880, was put to work that same evening:

About midnight, just as I was beginning to feel a bit lonesome at my solitary desk, a head popped in at the rear door and I was told to come to

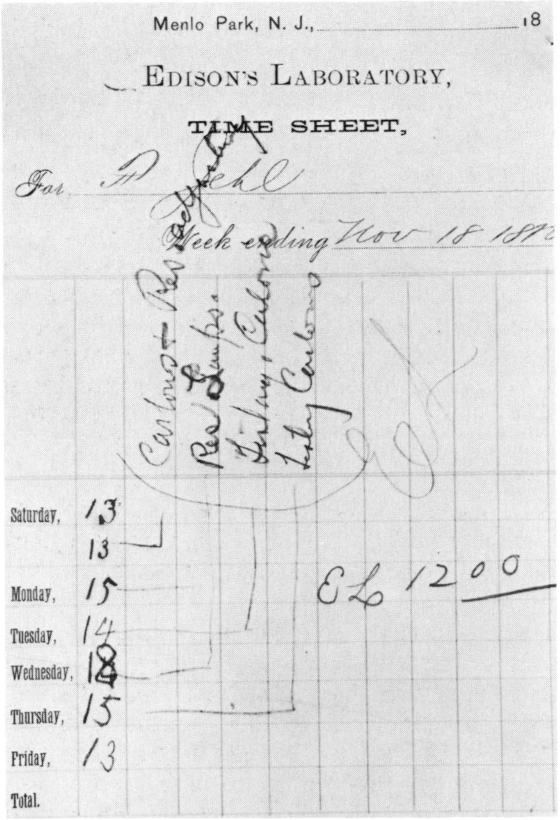
Menlo Park, N. J., ...18

Edison's Laboratory,

Time Sheet,

For F. Jehl

Week ending Nov 18 188

Saturday,	13
	13
Monday,	15
Tuesday,	14
Wednesday,	18
Thursday,	15
Friday,	13
Total.	

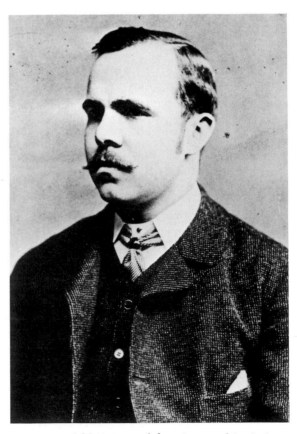

Figure 34. *Francis Jehl's time sheet for the week ending November 18, 1880, shows how workers assigned their time to specific activities so that it could be charged to particular projects.*

Figure 35. *Jehl, a young laboratory assistant at Menlo Park, later became the first caretaker of Henry Ford's Greenfield Village reconstruction of the laboratory.*

Figure 36. This engraving shows the second floor of the main laboratory building from the vantage point of where the organ stood.

the laboratory. There, on the second floor, way back near the pipe organ sat Edison, and scattered around were the night-working assistants and helpers; all, however, within convenient reaching distance of hamper baskets loaded with good things to eat that had been brought from Woodward's—not a cold lunch, but a hot dinner of flesh or fowl with vegetables, dessert and coffee. Hilarity came with the filling of stomachs, bantering and story telling were interlarded, until Edison arose, stretched, took a hitch at his waistband in sailor fashion and began to saunter away—the signal that dinner was over, and it was time to begin work again. This my first night meal in the laboratory was made the occasion for testing my fitness to join the brotherhood; I passed examination and was initiated with a few happy remarks by the great master himself.

Life in the laboratory was, as Clarke later described it, "strenuous but joyous . . . physically, mentally and emotionally," and they were working "frequently to the limit of human endurance." Elsewhere he describes what he was doing as "hot, messy, tedious work."

How could the laboratory operate under this sort of pressure? In part it was possible for the men to go on functioning because of the relief Edison provided, whether by closing down the lab and taking everybody fishing, or by offering those hot midnight snacks. Moreover, because the laboratory was the only game in town ("isolated as they were in the monotony of a rural neighborhood," to quote Clarke again), it was not unusual for people who were not on duty to wander in just to see what was going on. But mostly it was Edison's sense of how to work with other men. As John Ott put it, when he first met Edison, "He was as dirty as any of the other workmen, and not much better dressed than a tramp. But I immediately felt there was a great deal to him."

College-educated Charles Clarke used different words to describe the same thing: "The psychological atmosphere of the place partook of the inspirational, as a sort of reflex of Edison's genius, which was happily joined with a common sense and a commendable human nature that made him . . . 'one of the boys'. . . . Here breathed a little community of kindred spirits, all in young manhood, enthusiastic about their work, expectant of great results, often loudly explosive in word, emphatic in joke, and vigorous in action."

And then there is Francis Upton, who had persuaded Clarke to come to Menlo Park by telling him, "It is the chance of your life." Writing to his father, Upton reported, "I find my work very pleasant here and not much different from the time when I was a student. The strangest thing to me is the $12 that I get each Saturday, for my labor does not seem like work but like study and I enjoy it."

There were problems with this method of operation. As Upton wrote, "Mr. E. was sick for three days and during that time I had a fine chance to experiment to my satisfaction. One thing [that] is quite noticeable here [is] that the work is only a few days behind Mr. Edison, for when he was sick the shop was shut evenings as the work was wanting to keep the men busy." Henry Campbell, who joined the laboratory as a carpenter in October 1878, and may have had a slightly different viewpoint, recalled that "each and every man he had in his employ, simply carried out the request in their own special line . . . [doing] what Mr. Edison laid out for them to do. I know of no case, where anyone of us made new suggestions to Mr. Edison . . . but each one simply contributing to his work." Such a system can operate well as long as the "chief" is at hand, but it may not function at all when he isn't.

There is some indication, however, that Edison did encourage initiative in his workers, and not just in those who were obviously special, like Batchelor and Upton. Samuel Mott describes the use of competitiveness in his discussion of work done in April 1880, to make the Sprengel pump simpler and faster. Francis Jehl and Otto Moses had pursued different designs to this end. As Mott recorded it, "Francis put one wooden loop lamp on his pump in Dark room and got vacuum and heated up in two hours. Dr. Moses with two paper loop lamps got good vacuum in five hours but had not heated. From these first efforts it would be difficult to say which of the pumps were best and quickest for getting vacuum, but the pump in use by Francis is much the simpler, cheaper, and occupies less room. Boehm making a new pump slightly different from either just mentioned." Three days later he noted, "The single tube pump made by Boehm yesterday was started by Dr. Moses and has succeeded in getting a good pump vacuum in 17 minutes. . . ." And five days later he wrote, "Francis claims good pump vacuum on his pumps in six minutes."

Later that same summer, Wilson Howell was given the task of devising a means for insulating the wires that were to be buried underground for the forthcoming large-scale lighting demonstration. He recalled:

Mr. Edison sent me to his library and instructed me to read up on the subject of insulation, offering me the services of Dr. Moses to translate any French or German authorities which I wished to consult. After two weeks search, I came out of the library with a list of materials which we might try. I was given carte blanche to order these materials . . . and, within ten days, I had Dr. Moses' laboratory entirely taken up with small kettles in which I boiled up a variety of insulating compounds. . . . Of course there were many failures, the partial successes pointing the direction for better trials.

This is classic Edison experimental method, but done here with only the barest of guidelines and apparently with a minimum of oversight.

The men who comprised the work force at Menlo Park were a diverse group—the majority without any technical training, but a few with substantial experience or academic qualifications. Furthermore, the latter were more likely to be foreign-born or to have had experience abroad. Somehow they managed to get along with each other. Clarke had a room in a house next to Mrs. Jordan's boardinghouse, where he took his meals "along with other young, and some of them very lively, fellows of the laboratory." A few months later, as the number of workers increased, he moved, "joining the more sedate men in the Woodward home down near the station across the Pennsylvania Railroad tracks." But things didn't always work out well. Ludwig Boehm, the glass blower, left after little more than a year because he couldn't take the taunts of some of the other youths (Boehm was only 20 when he arrived). And no doubt there were other departures for similar reasons. Yet for the most part the force held together, and Edison was clearly the glue.

Figure 37. *The casual character of the workplace is evident in this 1879 photograph of Edison and his co-workers on the steps of the laboratory.*

Figure 38. *Sarah Jordan's boardinghouse provided a home and social center for Edison's employees.*

To have accomplished this with a dozen or so men during the first couple of years is an achievement; to have done so with 60 and more is truly remarkable. There is a hint of organizational structure in the way Edison treated Howell, and clearly he passed on some of the burdens of management through people like Upton, Batchelor, and Kruesi. Still, it was Edison who had the basic ideas and Edison who provided the inspiration. Whether he could have continued to operate a productive laboratory at Menlo Park after the electric light had been set on course, we can never know. But we do know that later, when he tried to enlarge the scale of operations at West Orange, something was missing. The number of personalities had become too large even for Edison's magnetic style of leadership.

In any case, the days of Menlo Park were about to come to a close. Clarke remembered that the results of the tests:

[M]ust have been immediately satisfactory to Edison and his backers in the Edison Electric Light Company. For in two or three days (on February 1st) Edison, boiling over with energy and enthusiasm, rushed up to me shouting: "come on, Clarke; pack up at once and come with me to

New York. We're going to begin business right off!" That very noon we walked into the "brownstone front" at 65 Fifth Avenue, which forthwith became the commercial offices of the Company. Edison rushed me up to the second floor, and said: "The Company has made you chief engineer. This is your office; the furniture will be here this afternoon. Furniture for your living room upstairs will be here, too—I want you on hand all the time."

And thus, on February 1st, my Edison-Menlo Park days ended. True, I used to go to the old place once in a while on business because I had two or three men there on special work; but I found the buildings almost deserted and nearly stripped of equipment and Edison no longer there. To me the spirit of the great inventor had left the place.

Bibliography

Some information about the workers at Menlo Park has been winnowed from the Edison biographies, four of which proved especially useful. Comments in Matthew Josephson, *Edison: A Biography* (New York: McGraw Hill, 1959), and Robert Conot, *A Streak of Luck: The Life and Legend of Thomas Alva Edison* (New York: Seaview Books, 1979), are generally referenced and have been employed with some feeling of confidence. Material from Mary Nerney, *Thomas A. Edison: A Modern Olympian* (New York: Harrison Smith and Robert Haas, 1934), and Francis Jehl, *Menlo Park Reminiscences*, 3 vols. (Dearborn, Mich.: The Edison Institute, 1934–41) lacks such checks and has therefore been used with greater caution, though these accounts tend to provide good anecdotal examples.

Of special value have been files on the "Edison Pioneers" in the Archives of Henry Ford Museum & Greenfield Village, at the Edison National Historic Site, and in the Archives Center of the Smithsonian's National Museum of American History.

The collections of the Edison National Historic Site include time sheets for the last half of 1878, four dates in 1879, and all of 1880, as well as a few payroll lists. These provided invaluable data for the graph. There are also a few letters, mainly from men seeking employment, and the journal kept by Samuel Mott. These are now available in Thomas E. Jeffrey, et al., eds., *Thomas A. Edison Papers: A Selective Microfilm Edition* Parts I and II (Frederick, Md.: University Publications of America, 1985 and 1987). Upton's letters are also preserved at the Edison Site, though there are copies in the Hammer Collection of the National Museum of American History's Archives Center.

As mentioned in the text, testimony at interference hearings has proven to be a unique source of detailed information about some of the lesser-known individuals. This material is found in the U. S. National Archives, Record Group 241. Those which proved most fruitful were Box 841, File 7943, *Boehm vs. Edison*, entries 18 and 19; Box 1621, File 12332, *Edison vs. Thomson*, entry 89; Box 1771, File 8195, *Edison vs. Maxim vs. Swan*, entry 195. A fully annotated version of this essay will be available from Division of Electricity and Modern Physics, National Museum of American History, Smithsonian Institution.

In assembling documents from these different locations the author gratefully acknowledges the assistance of Paul Israel and Leonard DeGraaf at the Thomas A. Edison Papers Project and of William S. Pretzer at Henry Ford Museum & Greenfield Village.

Machine Shop Culture and Menlo Park

Andre Millard

THIS ESSAY DESCRIBES the creation of a unique work culture at the Menlo Park Laboratory, a work culture of invention. The set of values and practices on which this culture was based stemmed from craft traditions of the pre-industrial era, traditions that stressed the skill of the worker and preserved the dignity and independence of his work. Thomas A. Edison had absorbed many of these values as an inventor and entrepreneur in the machine shops of the telegraph industry. When he established the Menlo Park Laboratory he deliberately instituted some of the practices he had found in the machine shops of Newark, New Jersey. This shop culture not only framed his method of inventing but also gave the experience of work in his laboratory its distinctive character.

Figure 39. *This famous photograph of 1880 shows Edison and his team on the second floor of the main laboratory building.*

Figure 40. *Taken around 1880, this photograph shows the brick machine shop at the far right and the office and library building to the left.*

The purpose of the "invention factory" at Menlo Park was to produce a steady stream of inventions that would lead to the development of new products. It was supposed to be half research laboratory and half factory. In trying to regularize a process that depended on the capricious and unpredictable powers of creativity, Edison was taking on a formidable challenge: if he was to achieve his goal of continual, systematic innovation he would have to reconcile the powerful forces of individual initiative on the one hand and capitalistic control on the other. Clearly he succeeded, for there can be no doubt that his idea of an invention factory was a commercial success. Edison's great achievement at Menlo Park was to create an informal work environment that fitted the peculiar needs of mass producing inventions. He made up his own work culture, and in doing so helped make the Menlo Park Laboratory unique in its creativity and its output of inventions.

Informal and democratic, the atmosphere in Edison's laboratory was a far cry from the insistence on discipline and order found in the contemporary factories of the textile industry or the large manufactories of heavy goods. In Newark and other industrial centers of New Jersey, men and women labored in an oppressive environment that was dominated by rules and regulations and the absolute power of the proprietors. At Menlo Park, by contrast, it was difficult to make out the boss in the group of laboratory workers, as the photographs of that time period show.

The unique atmosphere of Edison's laboratory was not merely the result of the peculiar quirks of the "Wizard"—the eccentric inventor figure romanticized in the press—but part of a work culture that stretched back to the craftsmen of Europe and their guilds and apprenticeships. Although the men of the Menlo Park Laboratory produced some of the most revolutionary new technologies of the 19th century, the tempo of work in Edison's invention factory reflected an ancient tradition of craftsmen and master.

Edison's first exposure to a craft culture came when he joined the ranks of telegraphers in the 1860s. This was the heyday of the "Knights of the Key," a group of skilled workers who shared a strong feeling of identity and a pride in their craft. Their telegraph keys maintained the communication required in the development of the American economy, for the telegraph industry provided an indispensible tool for business and government. Moreover the telegraph industry also became the avenue of advancement for several young inventors, Edison among them. Work in this industry proved uncertain, however, and Edison soon left the employ of the telegraph companies and set himself up as a designer of telegraph apparatus. He made up his experimental models in the machine shops of the telegraph industry in Boston and Newark, and soon made enough money to run his own shops in Newark. These early experiences in machine shops had a profound influence on his work methods and the organization of his laboratories. The shop culture that he encountered in Newark was duplicated in all of his laboratories from Menlo Park to West Orange.

This machine shop culture was centered on the well-honed skills of the worker. It valued the quality, rather than quantity of the final product. Craftsmen took pride in their work and were proud of their traditions, such as the "Blue Mondays" that followed a weekend of drinking, or the "blow out" that marked the passage from apprentice to journeyman. They often grouped together in social organizations and displayed a strong identification with the craft and their fellows. They developed their own vocabulary and way of dress which helped distinguish them from other workers. The brimless hat that Edison wears in one of the famous photographs of the Menlo Park lab was usually worn by glass blowers.

In the metalworking trades, the machine shop served as the locus of group activities and the practice of the craft. Each member of the shop served an apprenticeship which involved learning the trade and its special traditions. The craft culture functioned to give meaning and prestige to a skill by transferring a body of knowledge and a set of values. Once trained and socialized, the journeyman mechanic was free to work in the shops of his trade, traveling from place to place looking for work. He took his tools with him and often carried along special cutting tools that he had made up for specific jobs. Craftsmen honored work practices that gave the skilled worker—the journeyman—control over the pace and character of his work, as well as the important tools which he marked with his initials and kept in his own tool chest. (The tool box and marked tools that belonged to John Kruesi, for example, can be seen in the reconstructed laboratory at Greenfield Village, Fig. 43.)

These were the kind of artisans Edison came across in the machine shops of Newark. As the possessors of some very valuable skills they were accustomed to autonomy in the work place. They were hardly typical of the industrial workers in the 1870s. Unlike other craftsmen, such as shoemakers and weavers, who lost their skills in the forward march of industrialization, the mechanics and machine builders of the metal trades were given many opportunities to advance. The first machine shops were set up next to textile mills to repair machinery such as looms and steam engines. Before long, independent-minded mechanics established their own shops to repair, manufacture, and design new machinery. The machine shop became one of

Figure 41. The machine shop crew stepped outside for this 1878 photograph.

Figure 42. The view from foreman John Kruesi's desk in the Greenfield Village reconstruction shows the machine shop.

Figure 43. Kruesi's tool box—complete with marked tools and German engineering books—was donated by his family.

Figure 44. *The machine shop in the reconstructed Menlo Park complex is stocked primarily with tools and supplies donated by Thomas A. Edison in 1929.*

the most important centers of innovation during the 19th century.

Craftsmen in the machine shops stood at the forefront of a new industrial order. Their skills had helped usher in a new system of production, yet they still enjoyed some of the privileges of the craftsmen of the pre-industrial era—above all the control of the pace and timing of work. Artisans in machine shops and armories used a system of "inside contracting," by which skilled workers contracted with entrepreneurs to make the product. Although technically in the pay of the master of the shop, the craftsman could shape the steps of production to suit himself. He could work at his own pace—his was a tradition that did not accept the time clock—and subcontract with his fellows to help him.

In distinct contrast to the factories and mills of Newark, the machine shops gave individuals the opportunity to hire out the means of production, including tools, supplies, and labor. In some cases, proprietors of machine shops would rent out benches to independent craftsmen or inventors to make up their own apparatus. As a young inventor, Edison had enjoyed this privilege, renting the men and machinery (often on credit) to make up his experimental models. These practices fostered individual initiative and encouraged entrepreneurial activities on the part of the skilled mechanics who worked there: it is no accident that many important inventions in the new industries of electricity and communications came from the machine shop environment.

The men who worked in the Newark shops enjoyed a measure of freedom that was alien to the new industrial order. When Frederick Winslow Taylor, the father of what became known as Scientific Management, first entered machine shops in the late 19th century, he was horrified to discover that it was the men, and not the proprietor, who ran the operation. The workers, who would soon become Taylor's adversaries, controlled the work place and fashioned it to suit their own ends. The shops were places to socialize as well as work. Visitors were allowed easy access to the shops, whether they were men looking for work, boys looking for amusement, or amateur inventors looking for new ideas.

Figure 45. *Nineteenth-century machine shops were the site for much innovative work in machine building and design. Engaged in a relatively new trade, these machinists developed their own traditions of work.*

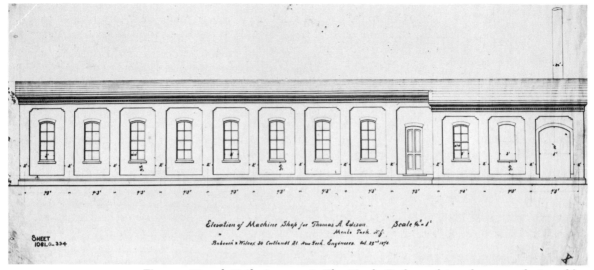

Figures 46 and 47 (facing page). *The Menlo Park machine shop was designed by the renowned engineering firm of Babcock & Wilcox of New York City, which also supplied the steam boiler.*

Although the craft system of apprenticeship was weak in the United States, the "tramp" system, whereby a young machinist roamed from shop to shop looking for work and broadening his experience, was a common practice in machine shops. Not only did itinerant mechanics extend their own educations by "tramping," but by carrying their ideas and tools from place to place they played an important role in helping diffuse technological know-how in the metalworking trades. There were neither factory fences nor guarded entrances in Edison's Newark shops. A constant procession of men came through the buildings, and not always to work; some lounged around, talking, smoking, and drinking. Work was stopped to buy newspapers from the urchins who haunted the shops. Drinking and card games were accompanied by the banter and laughter of a close-knit group of men.

This was exactly the kind of scene that might have confronted a visitor to Edison's Menlo Park laboratory: the informality, the conviviality of master and workers, and a general air of work unbounded by rules or timetables. Experimenting at the invention factory at Menlo Park was punctuated by gaming, practical jokes, and rowdy sing-songs at the organ placed at one end of the laboratory's second floor. The all-night experimental sessions, with their midnight feasts and hours of storytelling, became as important a part of the Edison myth as the inventions themselves.

Edison brought all the elements of the machine shop with him when he moved from Newark to Menlo Park in 1876. He set up a machine shop in the back of the first laboratory building established at Menlo Park. In 1878 he enlarged the laboratory complex, building a large machine shop behind the main lab building. Many of the physical items can be seen in the reconstructed laboratory in Greenfield Village, Dearborn: the work benches with vices and small tools to assemble equipment; the machine tools—lathes, planers, and drills—powered by a central steam engine connected to an overhead drive shaft; and the electrical equipment, such as relays, batteries, and testing apparatus. Edison also brought with him many ideas and practices that belonged to a craft culture. He continued the shop tradition of contracting with his men to make batches of telegraph equipment, or special one-off jobs that required extra skill in assembly. He maintained the openness and camaraderie he had enjoyed in the shops, and, most important, he created a sense of community in his Menlo Park work force.

The men at the Menlo Park Laboratory formed "a little community of kindred spirits, all in young manhood, enthusiastic about their work, expectant of great results," who worked, played, and lived together. They were bonded by their shared craft and by the special conditions under which they worked. Like European journeymen, the "muckers" (as they later called themselves at West Orange) lived a communal life and defended the borders of their fraternity. Only men Edison placed in charge of experiments, or who worked closely with the "Old Man," could join this select group at the top of the laboratory's hierarchy. Initiation into the brotherhood of muckers involved surviving a verbal harassment and proving one's worth in the daily challenges of experimenting.

The English working-class slang term "mucker" came from the verb "to muck in"—to pitch in, eating and working together, and to "muck about"—to fool about with little purpose other than amusement. "Mucking about" aptly describes work in the invention factory, where laughter and practical jokes abounded, many of which came from Edison's fertile mind and quick wit. The pranks and practical jokes (which often involved electrical equipment and painful electric shocks) were a regular part of working in the lab, and only the uninitiated or the naive would drop their guard when assisting Edison. The shop culture did not frown on practical jokes because they under-

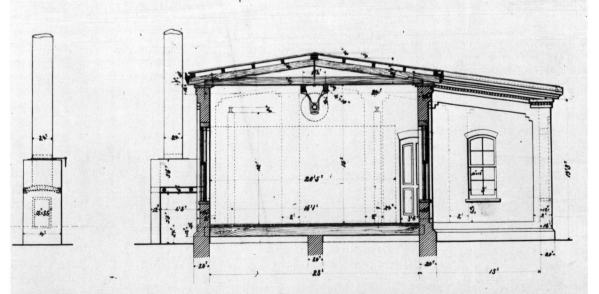

Scale ¼"=1'   Cross Section of Machine Shop for Thomas A. Edison.
Menlo Park. N.J.
Babcock & Wilcox 30 Cortlandt St. New York. Engineers. Oct. 23rd 1878.

lined the essential freedom of artisans in the work place. On the other hand, any self-respecting factory master would have been appalled at the loss of work time and risk of accident, and many were. Edison was more concerned with enjoying himself and harnessing the creative powers of his men.

Far from being sedate intellectual environments characterized by library quiet, Edison's labs were noisy, crowded places that often seemed on the point of uproar. As he told a new employee, the muckers did not work to any rules or regulations because they were trying to achieve something. That the invention factory was a pleasurable place to work in was part the heritage of the shop culture and part the result of Edison's intention to have it that way. His skill as a manager was to take some of the desirable elements of the

Figure 48. *This 1879 engraving from* Scientific American *shows Professor George F. Barker inspecting a bipolar dynamo being assembled at one end of the machine shop. Beyond the windows is the large C. H. Brown engine used to power the electric dynamos.*

Figure 49. The "Old Man" (in the middle, hatless, and minus a vest) sits with his men, who were young and robust, and ready to work long hours and risk many failures for their leader.

machine shop culture—motivation, initiative, and pride in work—and graft them onto an organization that he controlled.

Edison's well-known disregard for the 9 A.M. to 5 P.M. discipline of work was an example of the effective management style that came out of the artisan culture of the machine shops. The eccentric hours of working at the laboratory derived from the pre-industrial tradition of the shop. Craftsmen worked when they wanted and preferred variety to monotony in their work life. Much has been written about Edison's ability to go without sleep for long periods of time. His workers often stayed up all night with him. Work at the laboratory took no heed of the clock.

The experimental campaigns that went on for days made for good stories for the newspapers and gave the impression that invention at the lab went on night and day. What the public did not know was that Edison spent a great amount of his time taking cat naps, and after the all-night sessions the workers went to bed for the rest of the day. By adjusting the hours of work to suit his needs, Edison made the most of his workers' energy by keeping them at it when an experimental project was on the verge of a breakthrough and enthusiasm ran high.

The air of informality at the lab also helped to relieve the stress of producing a stream of innovations under a demanding timetable. Working for Edison at Menlo Park was a "strenuous but joyous life for all, physically, mentally and emotionally. We worked long night hours. . .frequently to the limit of human endurance," wrote Francis Jehl, who worked closely with Edison at Menlo Park. But many of his men could not stand the strain and left the laboratory. Tense and impatient during experimental campaigns, Edison kept the muckers under constant pressure, submerging them with work and often criticizing the results. The hours of storytelling provided an essential safety valve to the stress of work in the laboratory. The Menlo Park employees also relaxed by playing with the electric train that they built to run around the lab, or by going fishing in the local creeks. These activities reduced the intense pressure that often brought them to the edge of breakdown.

Edison was a successful manager because he gave his workers a sense of shared identification with his goals and transferred his enthusiasm to them. The men who worked at Menlo Park realized that they were part of a unique group of men who were making history. The daily challenges, the freedom to experiment and learn, participation in the varied social life of the muckers, and pride of association with the great "Wizard" and his laboratory, made work in the invention factory an exhilarating experience. One college-trained mucker commented that "my labor does not seem like work, but like study and I enjoy it." Yet it *was* work, and work of the most demanding kind: the muckers had to invent, improve, and perfect new technology under a hard master and a relentless series of deadlines. The machine shop culture provided the inspiration and the organization of the invention factory. It was a critical part of Edison's approach to the mass production of inventions. The vast number of experiments undertaken at his laboratory is testament to the success of his methods.

The shop culture not only served Edison by motivating his work force, it also became the means to blend together the wide range of skills and knowledge he had assembled in his lab. His employees at Menlo Park were as diverse as one can imagine, ranging from highly educated, proper young men such as Francis Upton, to semiliterate machinists of no fixed abode. Edison had to bring these men together into one focused operation without quenching the creativity and initiative that was at the heart of the process of innovation.

Despite his image as the "Wizard of Menlo Park," a lone, heroic inventor whose brilliant mind created magical inventions, Edison rarely worked alone. Innovation in his laboratories was a cooperative affair; the inventor's genius was translated into experimental models by the nimble fingers and experienced eyes of the skilled machinist. In his years in the telegraph industry Edison had grown accustomed to inventing in the machine shop environment, where the equipment, tools, and craftsmen were always at hand. Never too confident of his own manual dexterity, he always surrounded himself with skilled machinists like Charles Batchelor and John Ott, who could interpret his rough sketches and turn them into working, three-dimensional models in a very short time.

Edison worked closely with his machinists, overseeing the development of the model and changing it as he saw it take shape. It was in the process of altering an experimental model as it was being assembled that the germ

Figure 50. *The electric train, which ran on a circular track of about one-and-a-half miles, was a pioneering effort to apply the new power source of electricity. It also provided fast-paced excursions and transportation to a nearby fishing hole.*

of an invention sometimes emerged. In the inventor's own words: "Sometimes I get an idea and jot it down in the [laboratory note] book and sometimes I would get it while the machine was being made and change it and then note it in the book." These notebooks would be left on the experimental tables to be picked up whenever needed. Sometimes they were used by Edison to quickly record an idea, usually in the form of a rough sketch; at other times they were used to note the progress and outcome of an experiment after it had been completed. A page torn out of the book, bearing Edison's rapid pencil strokes, would be passed to one of the machinists on his staff, beginning the long road to an invention.

The tradition of the shop culture was learning by doing; in the Menlo Park lab it was inventing by doing: altering the experimental model over and over to try out new ideas. Although invaluable to historians and patent lawyers, the laboratory notebooks do not fully document the work of the lab, so that we have no record of the fruitful cooperation of experimenter and machinist as they pored over an experimental device in their efforts to make it work, and then make it work a bit better. Edison's comment that nothing that was any good would work by itself—"you got to make the damn thing work"—suggests the effort required to make experimental devices run. A visitor to Menlo Park could often find Edison at a workbench, shoulder to shoulder with Charles Batchelor, both intently working on a device, the sound of metal against metal punctuated by a stream of instructions and suggestions from Edison.

Charles Batchelor was Edison's closest associate during the Menlo Park years. The dark-bearded Englishman was typical of the men who made up the laboratory work force. He was a master of metalworking and an excellent draftsman. His patience and agile fingers were invaluable in the tricky job of putting the first fragile filaments in incandescent lamps. A recent immigrant, he had brought his craft skills across the Atlantic to the great industrial center of Newark, where he installed textile machines in the large Clark thread mills until he met Edison. Batchelor was one of many craftsmen who left the old world for the new during the latter half of the 19th century. The ports of New York and Newark formed a fulcrum of the Atlantic economy, an entry point for the Englishmen, Irishmen, Scots, Germans, Swiss, Frenchmen, and Scandinavians who went to work in the machine shops of Newark. They brought with them high degrees of skill and the traditions of a craft era. In Edison's shops one would probably have heard more German spoken than English, for many of the skilled machinists—including John Kruesi who ran the Menlo Park machine shop—were from central Europe.

The success of the invention factory idea depended on the skills of these craftsmen. Fast, flexible workers, accustomed to high standards of precision work, provided the foundation of the experimental teams that carried out the work of innovation. The fruitful cooperation between inventor and machinist, personified by the relationship of Edison and Batchelor, became the basis of the team approach to innovation that characterized work in Edison's laboratory. Edison made it clear later in his career that "the way to do it is to organize a gang of one good experimenter and two or three assistants, appropriate a definite sum yearly to keep it going...have every patent sent to them and let them experiment continuously." As this quotation indicates, the chief experimenter ran the group with the minimum of interference from Edison. He would outline the task and give some pointers, but he normally relied on the initiative of the experimenter. When one man asked him what to try next, Edison replied: "Don't ask me. If I knew I would try it myself."

The basic team comprised an experimenter who provided the leadership and machinists and other craftsmen who assisted the team leader. This experimental team was augmented by other machinists (and other skilled craftsmen) as needed. In many cases machinists would divide their time among different experimental groups, moving from team to team. The teams were free to draw on all the resources of the laboratory: supplies from the store room, scientific information from the library, pearls of wisdom from Edison, and the hands-on experience of the machinists and other craftsmen.

The muckers were encouraged to maintain communication among the teams because Edison knew that this was a vital part of the process of innovation. The key to one problem might be found in the experimental results generated in another project. The openness of the machine shops permitted the ready spread of information, not necessarily the written knowledge of science but the nonverbal information of ways of doing things, of techniques and prior experience. The ease of communication in the shops had helped bring forth a great number of important inventions. Edison ensured that this would also happen in the invention factory, where much of the fraternization of the muckers involved talk about their various experiments. During the late nights at Menlo Park, Edison sometimes led long, rambling discussions of the work at hand.

True to the democracy of the machine shop culture, Edison was always open to the suggestions and ideas of his men when it came to new experiments or inventions. Anyone was free to try a new idea. The machinists did more than act as Edison's hands—they filled in the details that were omitted in his fast-moving thoughts, and they applied their own expertise in the struggle to get the thing to work. They also took the initiative to modify devices and to experiment freely. Machinist and experimenter were partners in invention.

Edison had to make sure, however, that his men worked at several formidable tasks at the same time, for the invention factory had been built to produce results in many different fields. In addition to the uncertainties of experimental work, the logistics of running such an organization presented a severe challenge. How was the manager of such a complex undertaking going to manage several different experimental projects running simultaneously without losing control of the operation?

Edison's response was to maintain the machine shop tradition of personal leadership. Each day he would take a stroll about the laboratory buildings, going up to each man at the workbench, questioning him about what he had done, discussing the results, and deciding what to do next. This practice kept him well informed about the progress in each experimental project, giving him the opportunity to make his contribution in the form of suggestions for future experiments. It also preserved the element of personal contact that was the mark of the machine shops when the "Old Man" took a personal interest in the work of the artisans at the benches. Edison fitted easily into the role of the master of the shop and the ultimate authority on the direction of the experimental projects. Believing that the real measure of success is the number of experiments that can be crowded into 24 hours, Edison made his men work as hard and as long as he did. He was a demanding employer who did not suffer fools or poor workmanship gladly.

Yet he also enjoyed being one of the boys. An accessible, egalitarian boss, he liked to be among people, and the people whose company he enjoyed the most were his fellow experimenters. He dressed like one of his machinists and was normally as dirty as anyone in the machine shop. He made a habit of working alongside his men, and took part in the conversations and arguments that accompanied the experimenting, and the teasing and hazing of muckers, sometimes to the point of harassment. Partly this was fun and games and partly it was Edison's desire to test the mettle of his men. If they did not come up to scratch their future in the laboratory was uncertain. Edison never forgot that as owner-operator of the invention factory, it was his responsibility to ensure that it made a profit. He continually crossed the line from one of the boys to the often autocratic "Old Man" who hired and fired with the ease of a 19th-century factory owner. The shop culture had its uses and it made for an enjoyable work experience, but it often took second place to the demands of a calculated capitalism.

This emphasis is evident in the evolution of the accounting system used

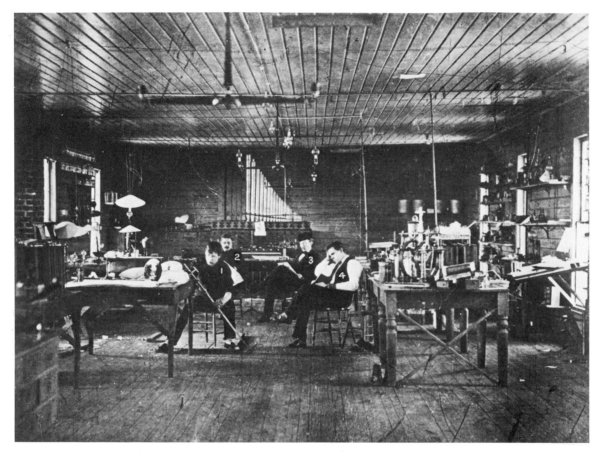

Figure 51. *Members of the staff relax by reading on the second floor of the laboratory. The incandescent light bulbs attached to the tops of the gas lamps are clearly visible.*

at the laboratory, a vital part of the Menlo Park operation that could not remain in the craft context. The essence of the mass production of inventions was an operation that supported several different experimental projects at the same time. This provided the economies of scale of a large work force and well-equipped laboratory, as well as the economies of scope that emerged when a result generated in one experiment proved useful in the successful conclusion of another. The great number of experiments carried out at Menlo Park and the diversity of the experimental projects undertaken were not the result of a chaotic mind or inflated ego, but part of a calculated plan to make the most of the resources accumulated under the roof of the invention factory. The traditional method of bookkeeping in the shop, which only counted the man hours and materials used on each job, was not adequate to the demands of running the invention factory. Edison had to keep up-to-date records of where his resources were being used to be able to make intelligent decisions about which project to terminate and which one to strengthen.

Keeping track of the time worked by a large group of men all paid on different scales was one of the managerial challenges facing the operator of an invention factory. The development of an accounting system to serve the requirements of continual, systematic innovation was part of the advanced industrial management that Edison grafted onto the craft culture of his laboratories. The accounts records of the Menlo Park facility form part of the gradual evolution of Edison's management techniques that started in the machine shops of Newark and culminated at the great laboratory at West Orange, New Jersey.

The early shop culture had no need for detailed accounts, for the masters worked on a small number of projects on a personal basis. The records kept by Edison and his partners in the Newark shops are predominantly accounts payable and receivable (a reflection of the complex interaction of the shops and their suppliers), and lists of all the costs involved in one specific job. Workers kept their own records of the hours they worked, and after confirmation by their foreman these totals became the basis of the payroll. The payroll itself, which recorded the weekly wages paid to each worker, was a separate sheet that each employee signed after he received his money. The weekly payrolls were sometimes copied into ledgers to form a permanent record of weekly labor costs—at this time the main concern of the proprietor of the shop.

The most valuable commodity of the invention factory was the labor expended by the skilled craftsmen and the experimenters. The allocation of this vital factor of production was the key both to success in the experimental work and to the profitable operation of the facility. There was an urgent need for a record of this allocation, not only for preparation of the payroll but also as the basic information for use in managing the operation.

The first steps toward this type of record keeping had been made in the early days of the Newark shops, when the workers' time was distributed over a small number of manufacturing jobs. Since most workers were paid by the hour it was easy to figure the time spent on each job. As more projects were undertaken at Menlo Park, and experimenting mixed with manufacture, special wage sheets were developed so that the workers' time could be distributed over several jobs. The sheets were in the form of a grid, with the days of the week on the vertical axis and the columns for each experimental project on the horizontal axis (see Fig. 53). After each day's work the employee noted how many hours he had spent on each job. Adding all the numbers on the horizontal axis gave the number of hours worked each day—the basis for pay computation—and the addition of the numbers under each experimental project gave an exact tally of the labor expended on each job.

It was then only a short step to systematically assign the labor costs over a large number of experimental projects. Allowance was made for repair work, set-up time, and general duties not connected to specific projects. (This was the first step in the long process of figuring the overhead costs of experimenting and incorporating them into the laboratory's accounts.) Separate lists were still made of labor costs and materials used for each job. This provided an exact total of cost but was a cumbersome indicator of the activity of the invention factory. Edison did not have the time to look at large books of lists. What he wanted was an idea of the weekly outlay on each experimental campaign on one easily read sheet of paper. The weekly distribution of labor costs fitted this requirement perfectly, and as labor costs constituted the bulk of the costs of experimenting this sheet of numbers gave an accurate (though underestimated) overview of the work of the lab. The costs of supplies, overhead, and depreciation were not included in these estimates.

In devising an accounting system that was probably more advanced than any other in use in machine shops or laboratories, Edison had the means to keep track of every experimental project. He used these sheets of numbers in the decisions about the business of innovation—what projects should be strengthened by the addition of more resources, which ones should be discontinued or cut back, and where to expect some significant results.

Edison applied the same ingenuity to his business affairs that he brought to his experiments. Faced with the problem of running a unique and complex operation at Menlo Park, he fashioned a solution out of what he remembered of the old and what he expected of the new. The machine shop culture was the means to blend many different skills into one highly motivated and powerful whole. He managed to obtain superhuman efforts out of a mixed bunch of talented men without losing control of the invention factory and its output. By devising a unique work culture that was controlled by the tools of modern capitalism, Edison managed to have the best of both worlds.

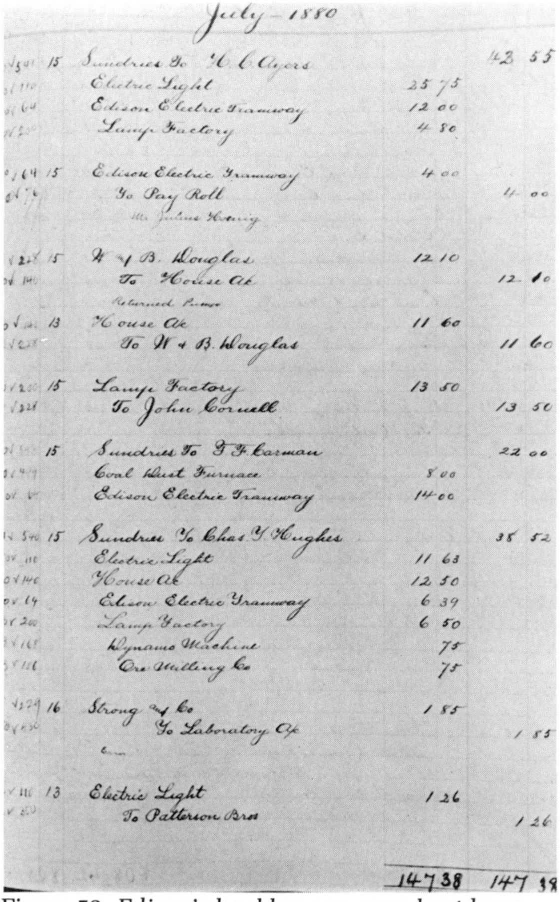

Figure 52. *Edison's bookkeepers were kept busy transferring the cost of wages, supplies, and services from time cards and invoices into several sets of ledgers.*

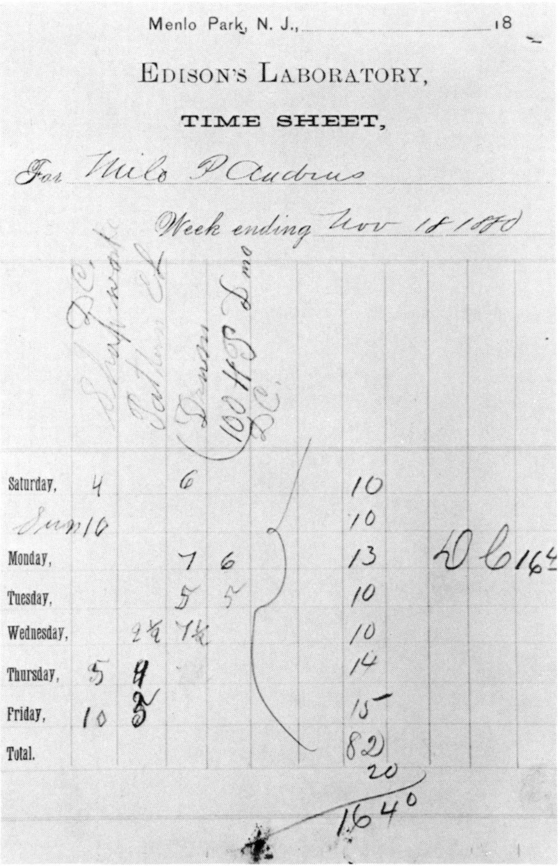

Figure 53. *The time card filled out by Milo Andrus illustrates the method of apportioning each part of a day's work to a specific project.*

Bibliography

Much of the primary source material for the Menlo Park years has been microfilmed by the Thomas A. Edison Papers project, Thomas E. Jeffrey, et al., eds., *Thomas A. Edison Papers: A Selective Microfilm Edition, Parts I and II (1850–1886)* (Frederick, Md.: University Publications of America, 1985 and 1987). The accounts of the Newark shops and the Menlo Park Laboratory can be found in Part I (1850–1878) on reels 20–22.

My understanding of the machine shop culture comes from Monte A. Calvert, *The Mechanical Engineer in America, 1830–1910* (Baltimore, Md.: The Johns Hopkins University Press, 1967), E. P. Thompson, *The Making of the English Working Class* (New York: Vintage Books, 1963), W. J. Rorabaugh, *The Craft Apprentice* (New York: Oxford, 1986), and Herbert Gutman, *Work, Culture and Society in Industrializing America* (New York: Vintage Books, 1976).

Useful studies of the craft culture of individual groups of workers include David Bensman, *The Practice of Solidarity: American Hat Finishers in the 19th Century* (Urbana: University of Illinois Press, 1905), Edwin Gabler, *The American Telegrapher: A Social History, 1860–1900* (New Brunswick, N.J.: Rutgers University Press, 1988), and William S. Pretzer, "The Printers of Washington, D.C., 1800–1880: Work Culture, Technology, and Trade Unionism" (Ph.D. dissertation, Northern Illinois University, 1986).

Telegraphy and Edison's Invention Factory

Paul Israel

THOMAS ALVA EDISON has rightly been seen as a transitional figure between the lone heroic inventors of the 19th century and the industrial research scientists of the 20th century. His new "invention factory" at Menlo Park, New Jersey, was itself a product of changes in the industrial context that were to foster the science-based industrial research laboratory of this century. Undergirding this new context was an emerging corporate culture that relied less on the invisible hand of the market and more on what historian Alfred Chandler has called the "visible hand" of modern management. The telegraph industry in which Edison's work had been centered was one of the first to move in this direction by establishing a centralized management structure to coordinate and control wide-flung operating units.

Figure 54. *Edison gained valuable experience in 1868 while using the facilities of Charles Williams, telegraph instrument manufacturer, located in this building on Court Street in Boston.*

Figure 55. *This photograph of Edison's shop on Ward Street in Newark was taken around 1872.*

As a telegraph inventor, Edison was strongly affected by the new strategies that managers of these firms developed to ensure their long-term growth and stability, and he established the Menlo Park Laboratory in direct response to Western Union's attempts to control technology as part of its strategy to maintain a dominant market position. While the emergence of this new corporate culture influenced Edison's decision to build his laboratory, and subsequently played an important part in his work on electric lighting, both the design of his laboratory and the style of his inventive work there derived from his experiences as a telegraph inventor. By placing the laboratory's beginnings and the work undertaken there within the context of Edison's early telegraph career, we gain insight into an important transitional period during which individual invention began to give way to collaborative efforts directed toward corporate goals.

While working on telegraph inventions, Edison relied on the resources of the precision machine shops of telegraph and electrical manufacturers. These shops were the central institution of telegraph invention, providing both a meeting ground for the exchange of ideas and facilities for the experimental development of instruments. Many telegraph inventors were owners or employees of machine shops or had working arrangements to use their facilities. While living in Boston in 1868, at the beginning of his inventive career, Edison found space in the machine shop of Charles Williams, Jr., who also provided the well-known telegraph inventor Moses G. Farmer with space for a small laboratory. Later Alexander Graham Bell worked on his telephone in Williams's shop, which specialized in producing experimental devices and employed skilled machinists to work with inventors. After moving to New Jersey in 1869, Edison worked briefly in the shop of noted electrical manufacturer Leverett Bradley before opening his own shop in Newark in 1870.

Throughout his career as a telegraph inventor, Edison sought to ensure himself access to machine shop facilities. Edison early learned the disadvantages of not having ready access to such facilities. Finding himself greatly delayed while conducting experimental tests of one of his inventions upon first coming to New York, Edison explained to his Boston backer, E. Baker Welch, that he was "awaiting the alteration of my instruments which on account of piling up of jobs at the instrument makers have been delayed..." [May 8, 1869]. Subsequently, Edison's contracts for inventive work incorporated provisions that enabled him to set up manufacturing shops or gave him access to such facilities.

Edison's first important contracts with a major telegraph firm, the Gold and Stock Telegraph Company of New York, which contracted with him to develop printing telegraphs and other new instruments for use in its market reporting and financial news telegraph services, included provisions that enabled him to establish his first shop, the Newark Telegraph Works, early in 1870. Soon thereafter, he also established the American Telegraph Works with funds from investors who later employed Edison's system of automatic telegraphy to compete with Western Union.* Between 1870 and 1875, Edison operated several telegraph manufacturing shops, which not only provided resources for his experimental work, but also gave him a steady income as the principal source of instruments for the Gold and Stock Telegraph Company and the Automatic Telegraph Company of New York. Although Edison was forced to devote his energies to manufacturing as well as invention, he left much of the day-to-day operations to partners in these enterprises.

The desire of telegraph companies to exert greater control over new technology soon provided Edison with new contracts from Gold and Stock and Automatic Telegraph by which he received annual salaries from both firms as a contract inventor and consulting electrician. He also attracted the attention of Western Union, which acquired control of Gold and Stock in 1871. Western Union also controlled the Stearns duplex, the first commercially successful system for transmitting simultaneously in two directions over a single wire, which it obtained in 1872. When the Stearns duplex provided an unexpected and important competitive advantage for Western Union, company president William Orton became "apprehensive that processes for working Duplex would be devised which would successfully evade [the] patents." In

---

* Automatic telegraphs employed automatic machinery in the high-speed transmission of messages. Operators using specially designed perforating machines punched holes representing the dots and dashes of Morse code onto a strip of paper and then fed this strip into a transmitter that passed an electrical contact over the perforations, causing the circuit to close intermittently and transmit signals. A receiver at the other end of the line recorded the signal by means of electrochemical decomposition on specially treated paper or by means of an ink-recorder. Edison's system, developed in the early 1870s, used electrochemical recorders because these could achieve higher speeds than ink-recorders.

Figure 56. *Edison sold the rights to his printing telegraphs, such as the early design shown here, to finance his inventing.*

his evolving strategy to control the new technology Orton commissioned Edison "to invent as many processes as possible [in order] to anticipate other inventors in new modes and also to patent as many combinations as possible" [December 2, 1872, letter to Joseph B. Stearns].

As Edison later testified, Orton engaged him to invent duplexes "as an insurance against other parties using them—other lines." In return for his efforts, Edison sought equipment and help from the company's machine shop, operated by another prominent telegraph inventor, George M. Phelps. Edison's work for Western Union produced unexpected results for the company when he developed the quadruplex telegraph, which enabled four messages to be simultaneously sent, two in each direction, over a single wire. Western Union's failure to secure a formal agreement with Edison for control of any inventions he developed with Western Union support later led to a protracted court battle with Jay Gould's Atlantic and Pacific Telegraph Company, which bought rights to the quadruplex from Edison. These legal difficulties gave Orton a new appreciation of the need to establish more formal relationships with inventors working for the company.

Edison's support from these telegraph companies enabled him to build improved laboratory facilities. Using money from his contracts with Gold and Stock and Automatic Telegraph and the savings resulting from his use of Western Union's shop and facilities, he was able to build a laboratory that by the end of 1873 he described as having "every conceivable variety of Electric Apparatus, and any quantity of Chemicals for experimentation" [December 1, 1873, letter to Charles E. Buell]. This was one of a number of laboratories that Edison established in connection with his various manufac-

Figure 57. *One of the most important items manufactured in Edison's Newark shop, the Universal Gold and Stock Ticker was an important source of income for him.*

turing shops. He later equipped a larger laboratory in 1874, which included, besides electrical equipment, such devices as a microscope, a spectroscope, and an air pump, as well as a more extensively equipped chemical laboratory. When Edison turned his full attention to inventing in May 1875, he separated his laboratory from the manufacturing operations (although he kept it in the same building), selling his rights in the firm to his partner Joseph Murray. The center of the new laboratory continued to be the experimental machine shop, where Edison employed three or four men who worked as full-time machinists.

A new contract with Western Union, entered into on December 14, 1875, enabled Edison to build his famous laboratory at Menlo Park, New Jersey. In exchange for his patents, Western Union agreed to give Edison access to its facilities, pay his patent expenses, and also provide up to $200 per week for experimental expenses. Western Union was prompted to do this by the fact that several other inventors were working on multiple telegraph systems employing tuning forks or other devices to send several frequencies at a time over a line. Fear of losing the competitive advantage it enjoyed in multiple telegraphy through control of the Stearns' duplex and Edison's quadruplex led the company to offer this additional support.

As in Edison's Newark laboratories, the machine shop played a central role in the inventive work. When faced with a strained budget due to the construction and operating costs of his new laboratory, Edison again turned to Western Union. Under the terms of a new agreement signed on March 22, 1877, at the inventor's initiative, Orton hoped to take advantage of the laboratory and its staff, which further enhanced Edison's proven abilities to turn out new inventions with startling regularity. Orton therefore agreed that Western Union would provide $100 per week for the ''payment of laboratory expenses incurred in perfecting inventions applicable to land lines of telegraph or cables within the United States.'' This money was used by Edison to support his mechanical department, the precision machine shop he adapted from his telegraph manufactories.

By adapting the machine shop solely to inventive work Edison and his assistants could rapidly construct, test, and alter experimental devices, thus increasing the rate at which inventions were developed. In this way the laboratory became a true invention factory. The best description of the machine shop at this time is found in Edison's letter to Orton asking for financial support for it:

> I have in the Laboratory a machine shop run by a 5 horse power engine, the machinery is of the finest description. I employ three workmen, two of whom have been in my employ for five years and have much experience. I have also two assistants who have been with me 5 and 7 years respectively both of which are very expert.

In the same letter, Edison stated that ''the cost of running the machine Shop including Coal Kerosene & Labor is about 15 per day or 100 per week. . .'' [Winter 1877]. A later photograph shows the shop filled with precision metalworking machines similar to those Edison employed as a telegraph manufacturer. Much of the machinery, as well as the men to run it, came from Edison's Newark shops.

Besides retaining the machine shop as the center of laboratory operations, Edison also drew on his telegraph experience when defining problems and seeking their solution. Edison's major work at Menlo Park—the telephone, the phonograph, and electric lighting—was directly influenced by his experience as a telegraph inventor.

Edison's first significant invention at Menlo Park was the carbon button telephone transmitter. He began working on this device after Alexander Graham Bell announced that his acoustic telegraph experiments had led him to invent a device he called the telephone which transmitted the human voice. Edison undertook his own telephone inventions in an attempt to provide Western Union with a means of getting around the Bell patent. The weak point in Bell's telephone system was his transmitter and Edison focused his efforts on improving this component. In Bell's transmitter sound waves vibrated a permanent magnet, which induced a varying current in the coils of an electromagnet; this current was transmitted through the line and used by the receiver to reproduce the signal as sound waves. The weak current set up by this transmitting device limited the distance over which it could be used.

In seeking a solution to this problem, Edison turned to a phenomenon he had observed in earlier experiments on cable telegraphs. After returning from England in 1873, where he had attempted unsuccessfully to transmit over a submarine cable using his automatic telegraph, Edison sought to increase his knowledge of cable telegraphy. In order to do so he built artificial cables on the tables in his laboratory. In one of these artificial cables he attempted to use the high resistance of carbon to replicate the resistance found

Figure 58. *The original Menlo Park machine shop, located at the rear of the first floor of the laboratory, was filled with the precision metalworking machines Edison had employed as a telegraph manufacturer in Newark.*

in a submarine cable. Unexpectedly, he discovered that any movement of the table caused by bumping or by vibrations and loud noises from machinery in the shop caused the resistance of the carbon to vary. Although this reaction was unsatisfactory for his cable experiments, it later proved to be the answer to the transmitter problem. In the Edison transmitter the pressure of sound waves against the carbon button varied a current flowing through it and thus regulated the strong current sent over the line, where the receiver converted it again into sound.

Edison's other major contribution to telephone technology also derived from a device developed in his early telegraph experiments. While competing with Bell in the British telephone market in 1878, Edison found it necessary to get around Bell's receiver patent. Several years earlier, while attempting to improve the electrochemical recorder of his automatic telegraph system, Edison had discovered that the friction of his metal stylus varied with the current flowing between it and the coated surface of his recording paper. This device, which Edison called an electromotograph, converted electrical signals into mechanical action, and Edison sought to use it in place of electromagnets in telegraph relays. After making extensive experiments he found that the surface of moistened chalk best exhibited these properties, and in 1878 he used an electromotograph employing a chalk drum (kept moist by being turned in a chemical solution) in the receiver of his new loud-speaking telephone. The chalk drum was fragile, however, and the operator had to continue turning the crank in order to keep the chalk moist. After the companies consolidated, the electromotograph-receiver was replaced by the Bell receiver, which was used in combination with Edison's superior transmitter.

Edison's most remarkable invention at Menlo Park, the phonograph, also drew on his experience with telegraph-recording systems, but within a new context of telephone transmission that he and other inventors were creating. Two problems related to telephone transmission concerned Edison during the experiments leading to the phonograph. First, he needed to devise a repeater in order to make long-distance transmission possible. Second, he and many others initially conceived the telephone within a telegraph context and thought that some means of recording messages would be important. From Edison's notebooks we know that while working on the telephone in 1877 he sketched a couple of devices for recording telephone messages that were very like recorders he had used in his automatic telegraph system. He also sought to devise a combination telephone repeater and recorder.

Edison's desire to record a transmission so that it could be played back in order to be accurately copied derived from one of his earliest telegraph inventions. The experimental telephone repeater recorded by indenting on a paper tape similar to a Morse embossing register. Edison had employed such a register as a practice telegraph operator's instrument that recorded and played back messages at a slower speed as early as 1865. For several weeks Edison's designs retained this telegraphic form of recording stylus and paper tape, but as he began to envision a recording device independent of the telephone he began to propose designs employing cylinders or discs. Edison called his new invention the phonograph and his most successful design was a cylinder on which he placed tin foil to act as the recording medium. The paper tape design was never used, although Edison did develop a disc instrument for the British market.

The phonograph demonstrated Edison's remarkable use of analogy for conceptualizing a problem. His success, however, often depended on moving away from his earlier conceptualization. This was most evident in Edison's research leading to the development of his system of electric lighting, which was greatly influenced by analogies with telegraphy. Edison's work on telegraphy and related inventions in telephony largely defined the limits to his knowledge of electrical technology at the beginning of his electric light experiments. Unlike some of his contemporaries, Edison did not have any experience with arc lights and dynamos. Yet he confidently predicted that he could solve the problem and after only a few days of experiment announced that he had "struck a bonanza" [September 13, 1878, letter to William Wallace].

Edison's confidence derived from his work in telegraphy. His solution to solving the problem of "subdividing the light"* was to use a series of relays and circuit breakers not unlike those he used in developing multiplex

---

* At the time Edison began his work, carbon arc lights were being used for outdoor lighting and for large interiors. The light of the carbon arc was too strong for small interior spaces, however, and electrical experts considered the problem of indoor lighting to be one of subdividing the light to make it comparable to that of gas lamps. They also recognized that such a subdivision required lamps employing incandescent rather than carbon arcs.

Figure 59. *The new enlarged machine shop, far left, reflected Edison's need for large, complex machinery, while the combined library and office, far right, reflected his increasing reliance on outside sources of information and financing. Note the electric train car next to the laboratory building.*

telegraphs. He believed that the problem lay in developing a regulator to prevent an incandescent burner (he was then using platinum and related metals) from melting. His long experience with making and breaking electric circuits in multiple telegraphy led him to assume that such devices would provide him with the solution to the light problem. The regulators Edison adopted for his lamps used the expansion of a metal wire or a column of air heated by the electric current to trip a switch, thereby breaking the circuit providing current to the incandescing element and preventing it from being destroyed by overheating.

Because Edison began his electric light research by relying on analogy from another field—telegraphy—his initial work focused on the lamp. He devoted his attention to varying the design of regulators and filaments in the same fertile manner that he varied the mechanical movements of printing telegraphs and the relays and circuits of multiplex telegraphs. Edison focused his work on the lamp because he believed that one of the existing generators could easily meet the requirements of an incandescent lighting system. After a month of experimentation, however, he began to realize the limitations of existing generators for his projected system of electric lighting, and he then attempted to design his own.

In order to understand the unusual design of Edison's first electric generator we must again return to his work on multiple telegraphy. Almost all of Edison's previous experience with electrical currents was derived from his work with telegraph batteries. His understanding of generators was limited to their relationship with electric motors like those used in driving the mechanisms of his printing telegraphs. Motors can be made to serve as generators (and vice versa) by altering their circuits, and Edison devised several such motors that grew out of his work on acoustic telegraphy. Acoustic telegraphs were a form of multiplex in which tuning forks were used to send and receive several messages of different frequencies over a wire at the same time. While conducting experiments on this type of telegraph system, Edison conceived the idea that the vibration of a magnetized tuning fork in combination with the electromagnets of a motor would increase the amount of available electric current. The similarity of motor and generator design made the tuning fork generator an obvious starting point for Edison. Once again, his telegraph experience helped provide intellectual resources for the electric lighting work.

Edison's tuning fork generator did not in fact work and he quickly recognized that he possessed insufficient knowledge of generator technology. Edison and his assistants therefore began a series of investigations into electromagnetism and generator design. They studied the existing designs, and in several cases actually acquired the machines, such as a Gramme dynamo, rather than relying on patents and other technical literature. In these experiments, Edison's familiarity with telegraph technology again led him to derive

Figure 60. *This 1878 photograph of the laboratory shows some of the sophisticated equipment and wiring used in Edison's experiments.*

Figure 61. *The elegance of the New York City headquarters of the Western Union Telegraph Company reflected the company's wealth and influence.*

Figure 62. *The main laboratory building as it appeared in 1879, when Edison was concentrating on the electric lighting project.*

an analogy, this time between the current in a generator's armature and the current produced by batteries, as a means to further his understanding of its design. For example, one notebook entry reads:

> *In Gramme armature the top half has a current running in one direction and bottom half a current in the other direction these must meet at O O [in the diagram] and the product is same as two batteries for quantity.*

And drawings of armatures often expressed the circuit as a set of batteries.

While these analogies with telegraphy provided a starting point for Edison's electric lighting research, the difficulties encountered in finding a solution to the problem forced him to move beyond the knowledge gained during his years of telegraph experience. In order to reach a deeper understanding of the problems facing him, Edison undertook a rigorous—some would say scientific—series of investigations into the properties of the materials he was using in his experiments.

This new spirit of investigation was most evident in the lamp experiments begun during January 1879. Instead of constructing numerous prototypes as before, Edison now began to observe the behavior of platinum and other metals under the conditions required for incandescence. In studying the effects of incandescence on platinum and other metals, Edison used an instrument that was not new but had heretofore seen little service—a microscope. In this way he discovered that the metal seemed to absorb gases during heating. To protect the metal Edison decided to place it in a glass container and evacuate the air with a vacuum pump, an instrument that had primarily been applied to scientific research rather than to technical problems.

Edison directed his efforts toward fundamental research so that he could gain a better understanding of what the optimal design of components for his electric lighting system should be. He was little interested in understanding why certain conditions had the effects they did. Instead he wanted to know what happened to particular materials or apparatus under those conditions. Edison and his staff also often referred to the basic scientific works in the laboratory library. In determining the appropriate materials to use for filaments, for example, he turned to such works as Henry Watts's *Dictionary of Chemistry*. Edison's experience with chemical texts and reference books dated back to his earliest work as a telegraph inventor when he consulted such sources as the *Proceedings of the Royal Society of London* and Michael Faraday's *Experimental Researches in Electricity*. A careful examination of the notebooks kept by Edison and his associates through the period of the electric light research, however, indicates that the electric lighting experiments prompted a series of careful and systematic literature searches in scientific and technical works that were far different in character from the patent searches or occasional references to individual works typical of Edison's earlier practice. Edison's interest in these studies, like his experiments on electromagnetism and filament materials, remained focused on what they could tell him about the actions of materials and forces under particular con-

Figure 63. *Edison's reliance on the machine shop is suggested by the number of machinists and helpers in this photograph, taken sometime in 1880.*

ditions as they pertained to his inventive work. While he could be temporarily diverted by investigations of unknown phenomena, he was first and foremost an inventor, not a scientist.

Because the light research required greater attention to scientific information and to complex calculations of data than had been necessary before, Edison hired a new type of assistant represented by college-educated Francis Upton, a graduate of Bowdoin College, Maine, who had done postgraduate work both at Princeton and under the aegis of Hermann von Helmholtz, a leading German physicist. On the one hand, Upton's academic training provided a sophistication in physical theory and scientific method that had been largely absent from Menlo Park prior to his joining the laboratory in December 1878. On the other hand, Upton claimed that he learned more about electricity from Edison than he had in his scientific coursework. Upton came to play a key role by calculating the technical and economic parameters of the system.

Because the design of components such as lamps and generators remained uppermost in Edison's mind, the machine shop continued to be fundamental to his inventive approach even as he refined his research techniques. The shop facilities and the large staff of the laboratory ultimately proved to be Edison's greatest advantage over his rivals. Edison's shop facilities enabled him to convert what he learned from his investigations into electromagnetism, filament materials, and vacuum technology into new designs for dynamos and lamps. He continued to rely on the skilled machinists who worked in the machine shop—headed by John Kruesi, who had been with Edison since Newark—to produce experimental devices for testing and redesign. The laboratory staff was headed by Edison's chief experimenter since 1873, Charles Batchelor, who was also a skilled mechanic and, like Edison, was self-taught in electricity. Given his large staff, Edison could work on several components at once and progress rapidly in designing his system.

Edison could call on such resources because of his earlier success as a telegraph inventor. Funds from his telegraph inventions provided the resources necessary to establish the laboratory in 1876 and in December 1878 new monies for the electric light researches came from investors associated with Western Union. Financiers of Western Union like J. P. Morgan and William Vanderbilt, and company officials like president Norvin Green, who became the first president of the Edison Electric Light Company, were willing to back Edison's venture in electric lighting because of their experience in supporting his telegraph experiments. They were also impressed with the international reputation as an inventive wizard that he had gained following his development of the phonograph.

The rich financial resources these backers provided—amounting to over $200,000 in the two-and-a-half years of active research and development on the electric light—enabled Edison to enlarge the best private laboratory in the country. Using these funds he built a larger machine shop, as well as a combined office and library building, which he stocked with scientific and technical books and journals.

At Menlo Park Edison began forging a new mix between traditional shop-trained mechanics and university-educated engineers, as the complexity of problems in electric lighting and power created the need for engineers with more sophisticated training in mathematics and physics. As research on electric lighting moved to the development stage in 1880, the staff expanded from about a dozen to 30 or 40 men. Most of these continued to be shop trained, but a few, like Charles L. Clarke (Upton's schoolmate at Bowdoin College), were trained mechanical engineers. Edison himself recognized the need for university-educated engineers and supported the development of electrical engineering education in the 1880s. In his later laboratory at West Orange, he hired many such men. However, Edison also shifted the focus of his work at West Orange away from lighting and other electrical technologies as, by the 1890s, a physics-based electrical engineering profession altered these technologies in ways that made his shop-training and self-education in electricity obsolete.

The changes in electrical technology that led Edison away from the field also created changes in the manner in which invention took place at electrical companies such as Bell Telephone and General Electric, both of which occasionally employed Edison as a contract inventor. Much of the early inventive work in telephony occurred in telephone manufacturing shops, most of which were associated initially with telegraph manufacturing. After the Bell Company purchased the Western Electric Manufacturing Company from Western Union in 1882, this shop became a principal source not only of telephones but of telephone improvements. The company also established an electrical and patent department in 1880, which was placed under the direction of Thomas A. Watson, Bell's principal experimental assistant and formerly a machinist in Charles Williams's shop. The following year, William

Figure 64. *This patent model for Edison's electro-harmonic multiple telegraph illustrates the combination of electrical and acoustical elements that led Edison to develop both the phonograph and the telephone transmitter.*

Figure 65. *The patent model of a carbon transmitter developed by Edison in 1877 shows one stage in the development of a significantly improved telephone transmitter.*

Jacques, one of the first Ph.D. graduates of the new physics/electrical engineering program at Johns Hopkins University, replaced Watson as head of the department.

The choice of Jacques to head this department reflected the changes occurring in electrical engineering. These changes were also reflected in the history of an experimental shop set up by the company in 1883. Initially this shop was placed under the direction of Ezra T. Gilliland, a former telegraph operator and manufacturer and an associate of Edison. In 1884 this shop, now known as the Mechanical Department, was placed under the charge of Hammond V. Hayes, Bell's second Ph.D. scientist. Under Hayes the Mechanical Department began to focus on new problems of long-distance transmission that represented the Bell Company's long-term strategy to compete with Western Union following the expiration of the initial telephone patents. With the incorporation of the American Telephone and Telegraph Company in 1885, the Bell interests began to develop seriously the technical means to accomplish this. The problems of long-distance transmission were primarily electrical rather than mechanical, and required sophistication in mathematical analysis and advanced physics. New electrical engineering programs established in American colleges by physicists such as Henry Rowland, who saw telephony and electric lighting as important fields requiring advanced physics, supplied engineers to solve these problems. Long-distance telephone technology and the new challenge of radio and electronics led the company to create its renowned industrial research laboratory by 1925.

So also at General Electric. Besides Edison, the company at first relied on the work of Elihu Thompson, the inventor whose arc-light and alternating-current systems formed the other part of the merger that made up General Electric. Thomson, however, had much more formal scientific schooling than Edison; he continued to work on problems for General Electric throughout his life and came to identify himself as an electrical engineer. General Electric funded a small laboratory run by Thomson, but also relied on university-educated electrical engineers working in its Standardizing Laboratory and Calculating Departments to solve design problems and problems of long-distance transmission as it adopted alternating current systems. The expiration of General Electric's basic lamp patents at the beginning of the century, however, created new competitive pressures that led the company to establish the first American industrial research laboratory which employed chemists and physicists.

The changes wrought by Edison at Menlo Park helped lay the groundwork for modern industrial research, although they did not directly result in such new institutions. The support Edison received for his work in electric lighting and telecommunications at Menlo Park helped demonstrate the

Figure 66. *The Western Electric Company's circuit development laboratory, shown here in 1920, eventually became part of the Bell Telephone Laboratories.*

value of invention to industry, and showed that invention itself could be an industrial process. His work in electric lighting also helped create the new electrical engineering profession that would use mathematics and physics to solve new problems.

Edison's example was important in shaping these new institutional patterns and preparing the ground for modern industrial research. However, as companies such as General Electric and AT&T established new industrial research laboratories, which employed physicists and chemists as well as electrical engineers, they showed the effects of a cluster of influences. These included new competitive strategies, European (particularly German) academic and laboratory traditions, and a growing belief in America that basic scientific research would lead to new and better technology.

Bibliography

None of the standard works on Edison deal adequately with his career as a telegraph inventor. Reese V. Jenkins, et al., eds., *The Papers of Thomas A. Edison: Volume 1, The Making of an Inventor (February 1847–June 1873)* (Baltimore and London: The Johns Hopkins University Press, 1989) and Thomas E. Jeffrey, et al., eds., *Thomas A. Edison Papers: A Selective Microfilm Edition, Parts I and II (1850–1886)* (Frederick, Md.: University Publications of America, 1985 and 1987) contain most of the documents detailing Edison's inventive work discussed in this essay. Other material is found in the records of the Western Union Telegraph Company, Upper Saddle River, New Jersey. Paul Israel's Ph.D. dissertation, "From the Machine Shop to the Industrial Laboratory: Telegraphy and the Changing Context of American Invention, 1830–1920" (Rutgers University, 1989), is the first detailed examination of invention in the telegraph industry and of Edison's role as a key telegraph inventor. On the telephone see Reese V. Jenkins and Keith A. Nier, "A Record for Invention: Thomas Edison and His Papers," *IEEE Transactions on Education* E–27 (November 1984): 191–196, and George Prescott, *Bell's Electric Speaking Telephone* (New York: D. Appleton & Co., 1884; reprint Arno Press, 1972). The best description of Edison's inventive work on the phonograph is Charles Batchelor's 1906 reminiscence published as an appendix in Walter L. Welch, *Charles Batchelor: Edison's Chief Partner* (Syracuse, N.Y.: Syracuse University, 1972). On the electric light see Robert Friedel and Paul Israel, with Bernard S. Finn, *Edison's Electric Light: Biography of an Invention* (New Brunswick, N.J.: Rutgers University Press, 1986).

On early invention in the telephone industry and its relation to telegraphy see George David Smith, *The Anatomy of a Business Strategy: Bell, Western Electric, and the Origins of the American Telephone* (Baltimore and London: The Johns Hopkins University Press, 1985). On the emergence of modern industrial research at General Electric and AT&T see Leonard S. Reich, *The Making of American Industrial Research: Science and Business at GE and Bell, 1876–1926* (New York: Cambridge University Press, 1985). On Elihu Thomson see W. Bernard Carlson, "Elihu Thomson: Man of Many Facets," *IEEE Spectrum* 20 (October 1983): 72–75 and his dissertation, "Invention, Science, and Business: The Professional Career of Elihu Thomson, 1870–1900" (University of Pennsylvania, 1984).

Thinking and Doing at Menlo Park: Edison's Development of the Telephone, 1876–1878

W. Bernard Carlson and Michael E. Gorman

ASK ANY SCHOOLCHILD "who invented the telephone?" and you will invariably receive the answer "Alexander Graham Bell." This is certainly correct, for Bell was the first to invent a device for transmitting the human voice through an electrical wire. Yet who converted Bell's rudimentary instrument into a reliable device which could be used by anyone? While the typical schoolchild knows the name of this inventor, he or she probably does not realize that he played a major role in developing the telephone. Thomas Alva Edison was foremost in perfecting the carbon transmitter or microphone still found in many of today's telephones.

Edison worked on the telephone at Menlo Park from 1876 to 1878. Next to the electric light, the telephone was the largest and most significant project that he pursued at Menlo Park. It was the largest pre-electric light project in terms of the time Edison devoted to it. Amply supported by money from Western Union, Edison concentrated on the telephone from November 1876 to March 1878, a period of 17 months. The project is significant in that Edison's contributions were quickly taken up by the nascent telephone industry and put into commercial use by mid-1878. It is important too in that Edison perfected his method of invention while working on the telephone and he used this same research program in his quest for the electric light.

This essay recounts the process by which Edison developed the telephone, focusing on his efforts to perfect a powerful and reliable carbon transmitter in 1877. As Edison succinctly characterized this process in 1878, "I had to create new things and [overcome] many obscure defects in applying my principle." As this quote suggests, invention may be seen as involving two elements similar to what Edison called his "principle" and "new things." First, an inventor has a principle or mental model of how he or she thinks his or her creation should work. Second, an inventor uses "things" or devices to express his or her mental model in physical terms, and these devices will be called building blocks. As this essay will reveal, Edison had a distinctive set of building blocks and often borrowed these from established inventions to create a new invention.

In this way, the act of invention may be seen as the interplay of mental models and building blocks. In developing something new, an inventor may begin with a mental model. This model incorporates a general idea of how a device might work and an awareness of its potential significance. By manipulating and experimenting with a selection of building blocks, an inventor explores variations and changes. Eventually, insights from the building blocks may lead an inventor to modify his or her mental model. An invention may be said to be complete when an inventor feels the fit between the mental model and the building blocks device is close, and when he or she is able to convince others that the device matches their mental model and expectations.

In the case of the telephone, the mental model Edison used was the principle of variable resistance. From his previous telegraph research, Edison knew it was possible to convert sound waves into electrical signals by varying the resistance in an electrical circuit; if the sound waves could vary the resistance proportional to their amplitude, then they would produce a fluctuating electric current. Using another device, this fluctuating current in turn could be converted back into sound waves. Initially, Edison investigated this mental model by using an established telephone device—this was his first building block—but over time he added other blocks from previous inventions until he produced a telephone superior to Bell's. What is interesting is how mental models and building blocks are linked. It appears that for a time the configuration of the building blocks guided the process, yet, by the end of the process, Edison had succeeded in completely modifying both his mental model and his building blocks.

From Multiple Telegraph to Telephone, 1875–1876

Like Bell, Edison became familiar with the knowledge and devices needed for a telephone while working on an entirely different project, a multiple message telegraph. During the middle decades of the 19th century, inventors and businessmen perfected the electric telegraph and established it as a major form of rapid communication. By the mid-1870s, this technology was largely controlled by Western Union, one of the first giant corporations established in the United States. Yet in building a nationwide telegraph network, Western Union was hampered by a severe technical-economic problem: as the volume of messages grew, the cost and the complexity of the network grew even more quickly. Consequently, if one needed to send ten telegrams simultaneously from, say, New York to Cincinnati, one was faced with two choices. First, one could string ten wires between the two cities, with each message on a single wire. (Of course, stringing dozens of wires over hundreds of miles from one city to another required a substantial amount of capital.) Second, one could devise a system of high-speed telegraphy whereby each message was sent faster, perhaps in one-tenth the time of a typical telegram. Intent on solving the problem raised by the volume of messages, the managers of Western Union took a great interest in technological innovation and encouraged a number of inventors to create improved telegraph devices.

Among the inventors who responded to the challenges and opportunities of improving telegraphy was Thomas A. Edison. During the early 1870s he developed several multiple message telegraph systems—first a two-message system (duplex) and then a four-message system (quadruplex). Although Western Union did install several of Edison's inventions on their lines, it also encouraged him to patent as many different designs as possible so as to block parallel activities by other companies and inventors. Because Edison provided inventions which could be used both offensively and defensively, the president of Western Union, William Orton, arranged for the company to underwrite much of Edison's research. With this financial support, Edison was able to withdraw from his telegraph manufacturing enterprises in Newark and build a new laboratory at Menlo Park in the spring of 1876.

One of the first items Orton asked Edison to investigate in 1875 was an acoustic telegraph. Although Western Union had installed the Edison quadruplex on its lines, the company hoped to increase still further the number of messages that could be sent simultaneously. Several inventors (including Elisha Gray of Chicago and Alexander Graham Bell of Boston) thought that the next step would be to assign an acoustic tone to each message and convert that into a fluctuating current. This current could then be sent over a wire and received by a special relay with a reed tuned to vibrate at the frequency of the original tone. While the principle of sending and receiving one message using an acoustic signal had been demonstrated, no inventor had yet succeeded in sending and receiving several signals simultaneously.

To begin work on an acoustic telegraph, Orton asked Edison to study the Reis telephone (Fig. 67). Invented in Germany by Philip Reis in 1861, the

Figure 67. *This telephone was invented by Philip Reis in Germany in 1861. As one spoke or sang into the mouthpiece* **S**, *the sound waves vibrated diaphragm* **mm** *and caused the platinum needle* **ab** *to move in and out of the mercury* **s**. *This movement opened and closed the circuit, thus creating a fluctuating current in the circuit. At the receiver* **C**, *this current rapidly magnetized and demagnetized coil* **M**, *which in turn caused the needle* **TT'** *to vibrate and reproduce the original sound.*

Edison began his research on the telephone by having his assistant James Adams test a similar device in the fall of 1875. In these experiments, Adams substituted a variety of electrolytic solutions for the mercury in cup **s**.

Figure 68. *Bell's sketch for a liquid transmitter is dated March 1876. By speaking into the opening at the top of the container, one caused the diaphragm at the bottom to vibrate. Attached to the diaphragm was a needle which moved up and down in a cup of water. While in the Reis telephone the needle broke contact with the mercury, Bell had his needle stay in the water. By varying the surface area of the needle in contact with the liquid, Bell secured a change in the resistance and hence produced a fluctuating current.*

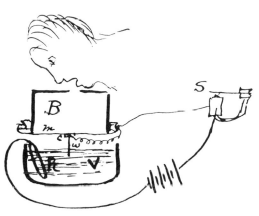

transmitter consisted of a membrane diaphragm connected to a needle in a small cup of mercury. If one sang or spoke into the diaphragm, the needle moved in and out of the mercury, thus opening and closing an electrical circuit and creating pulses of current. The receiver consisted of a long iron rod surrounded by a coil; the current pulses from the transmitter rapidly magnetized and demagnetized the coil and thus caused the rod to vibrate. Although Reis thought his invention might be able to transmit speech (hence the name telephone), he never succeeded in demonstrating this and instead concentrated on sending musical tones. Edison probably constructed a Reis telephone in the fall of 1875, but he soon set it aside to concentrate on developing an acoustic telegraph with more familiar telegraph instruments. Through the fall of 1875 and the spring of 1876, Edison was further distracted from studying acoustic telegraphy by his discovery of strange sparks, which he declared to be a new "etheric force."

As Edison pondered the "etheric force," Alexander Graham Bell was busy patenting his own telegraph inventions. Like Edison and Gray, Bell was fascinated by the possibility of developing a multiple-message acoustic telegraph. In pursuing this invention, Bell was actively encouraged by his future father-in-law, Gardiner Hubbard, a bitter opponent of Western Union, which he viewed as a monopolistic giant. Hubbard hoped to slay the giant by having Bell develop a multiple-message system that could then be used to create an alternative telegraph network. Although sympathetic to Hubbard's goal, Bell was soon excited by another vision. Drawing on his extensive work in teaching the deaf to speak, Bell saw in the acoustic telegraph the possibility of a speaking telegraph. During 1875, with the help of Thomas A. Watson, Bell conducted a number of promising experiments that confirmed his vision, and in February 1876 he filed a patent claiming the electrical transmission of speech.

To confirm his vision of the electrical transmission of speech, Bell initially used the principle of variable resistance. He devised a telephone in which the sound waves vibrated a thin diaphragm that was connected to a needle suspended in a cup filled with a liquid of high electrical resistance.

As this needle vibrated, the surface area of the needle in contact with the liquid changed, thus varying the resistance and strength of the current in the circuit (Fig. 68). To Edison, Bell's first telephone was not a particularly impressive device; in Edison's opinion, Bell had simply modified the Reis telephone by converting the make-and-break switch into a liquid rheostat. Liquid rheostats were hardly new to Edison, as he had used one in an 1873 patent for his duplex system (Fig. 69).

Although Bell had demonstrated voice transmission with his liquid transmitter, in the spring of 1876 he shifted to an entirely different principle, that of electromagnetic induction. Apparently, he decided that most liquids had too high a resistance and hence presented complications in getting a sufficiently strong electrical signal. Instead of connecting the vibrating diaphragm to a variable resistance device, Bell now linked it to the armature of an electromagnetic relay; as this armature vibrated in response to the sound vibrations, it induced a current in the relay coil (Figs. 70 and 71). Although this magneto design was simpler than the liquid transmitter, the currents generated even by shouting into the telephone were so weak that it was nearly impossible to hear the message in the receiver.

Through the summer of 1876, Bell gave several public demonstrations of his telephone, including one at the Philadelphia Centennial Exhibition. With the help of Hubbard, Bell apparently approached Western Union and offered to sell his telephone to the company for $100,000. During the fall, Orton, Hubbard, and Bell discussed the possible purchase of this invention, but eventually broke off negotiations. In all likelihood, Orton did not enjoy negotiating with Hubbard, an outspoken enemy of his company. But beyond this personal animosity, Orton seems to have decided that there was no need to buy Bell's patent because the telephone could be easily duplicated and improved by inventors already associated with Western Union. With talented experts such as Edison, Elisha Gray, and George M. Phelps, who worked in

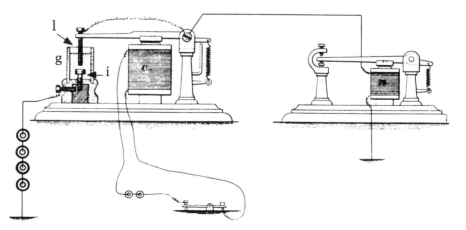

Figure 69. *The liquid rheostat that Edison patented in 1873 for use in his duplex telegraph is shown attached to the telegraph relay on the left. It consists of a cup* **g** *filled with either water or glycerine with contacts* **l** *and* **i**. *As* **l** *moves closer or further away from* **i**, *the resistance varies proportional to the distance that the current has to pass through the liquid.*

Figures 70 and 71. Reproductions of Bell's magneto telephone from 1876 (transmitter on the left). By speaking into the black cone-shaped mouthpiece of the transmitter, one caused a membrane diaphragm to vibrate. Attached to the diaphragm was a small iron armature. As this armature moved in response to the sound waves, it induced an electric current in the horizontal electromagnetic coil. The receiver consisted of an electromagnetic coil enclosed in an iron cylinder and covered with a thin metal diaphragm. The current from the transmitter created a magnetic field in the receiver coil, which, in turn, caused the diaphragm to vibrate, thus reproducing the sounds.

the electrical shops of Western Union, and were knowledgeable about the acoustic telegraph, it seemed far more sensible for Western Union to perfect its own telephone.

During the summer and fall of 1876, as Bell and Orton talked about a possible deal, Edison and his associates began to investigate the telephone at the Menlo Park Laboratory. Unimpressed with the magneto telephone, Edison drew on his experience with both liquid rheostats and the Reis telephone. Edison had his assistant, James Adams, undertake a series of tests with a modified Reis telephone. Instead of using mercury in the little cup, Adams experimented with drops of different liquids and then sponges, paper, or felting dipped in various electrolytic solutions. Unfortunately, the liquids quickly evaporated and Edison concluded that a high-resistance liquid was not suitable for a practical telephone. Yet these early experiments with a modified Reis instrument seem to have convinced Edison that a successful telephone would be based on the principle of variable resistance and not electromagnetic induction. These experiments established in his mind the mental model of variable resistance that would guide him through hundreds of experiments in 1877. Furthermore, intimacy with the Reis telephone provided Edison and his associates with a tangible building block for their experiments; throughout the next few months, whenever they were stymied, they would return to the Reis telephone to review their mental model and get ideas for new experiments.

Discovering and Perfecting the Carbon Transmitter, 1877
By January 1877, it had become clear that Western Union would not buy Bell's telephone and Edison began working on his own version in earnest. Abandoning the messy liquids, Edison and his associates decided instead to use a thin film of graphite or carbon which possessed a high resistance. They placed this film on various surfaces and then had a platinum needle connected to the speaking diaphragm make contact with the film. As the needle vibrated and moved across the film, the resistance of the circuit varied proportionately with the distance that the needle moved.

In designing this new transmitter, Edison's mental model of variable resistance provided not only an abstract principle to investigate but also a visual and tactile way of arranging the apparatus. Because he had secured the phenomenon of variable resistance previously in devices that moved needles in and out of high-resistance materials (as in both his liquid rheostat and the Reis telephone), Edison used this configuration in the carbon film telephone. Guided by this mental model, Edison initially decided that the way to improve the telephone was to experiment with how the needles made contact with the film, and he asked his master machinist John Kruesi to fashion several different transmitters, each with a delicate arrangement of needles and film surfaces (Fig. 72).

For the next several months Edison experimented with these different mechanical arrangements, uttering phrases such as ''Physicists and Sphynxes

Figure 72. *Edison developed this experimental transmitter in early 1877. It worked by having several platinum needles attached to a diaphragm move across a film of graphite mounted on the small dark block in the center of the device.*

[sic] in Majestical Mists'' into the crude transmitters to test for volume and articulation. However, he found that no matter how he arranged the needles, their movement distorted the sound waves and garbled the transmission. This difficulty forced Edison to rethink how he was using carbon in the telephone, and in April 1877 he introduced a new building block to his telephone by borrowing from an old invention. In 1873, while working on problems relating to undersea telegraph cables, Edison had needed a rheostat that simulated the resistance of a very long cable. Knowing that carbon (like other semiconductors) varied its resistance when placed under mechanical pressure, Edison had designed a rheostat made of glass tubes filled with fine particles of carbon. Because the carbon grains were squeezed into the tubes, they provided an enormous electrical resistance and one could adjust the resistance of this device by inserting or removing different plugs (Fig. 73). Unfortunately, this rheostat proved unsuitable for Edison's cable experiments because the resistance of carbon varied whenever the apparatus was bumped. However, Edison now took advantage of this limitation and applied it to the telephone transmitter. Reasoning analogously from this familiar device, Edison constructed a telephone in which the speaking diaphragm was connected to a small platinum contact which pressed directly on the carbon (Fig. 74). As one spoke into the diaphragm, the force or pressure placed on the carbon changed and varied the resistance.

Preliminary results with this new telephone quickly convinced Edison that carbon-under-pressure was an excellent building block. Ever anxious to reassure his supporters at Western Union, Edison told them in May 1877 that his telephone was ''more perfect than Bell's'' and that they ''need to have no alarm about Bell's monopoly as there are several things that he must discover before it [i.e., Bell's telephone] will be at all practical.''

Figure 73. *Edison's carbon rheostat of 1873. By inserting or removing metal plugs from the holes on the top of the rheostat, one could increase or decrease the number of glass tubes in the circuit and hence vary the resistance. This design closely resembles a standard resistance box; the major difference is that Edison used the carbon-filled glass tubes.*

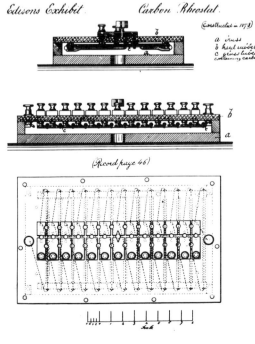

Figure 74. *This experimental carbon telephone transmitter was used by Edison in the spring of 1877. Carbon samples were placed between two flat electrodes inside the mouthpiece. The amount of pressure on the carbon could be varied by adjusting a long spring (now missing) that was fastened to the binding post to the left of the mouthpiece.*

Yet as Edison experimented with this new building block, he realized that what he wanted was a carbon compound that was very sensitive to mechanical force. Ideally, a small change in the pressure on the carbon should produce a large change in the resistance, thus amplifying the signal. Hence the task was to find a carbon compound with this electrical property, and Edison set his associate Charles Batchelor to work testing a wide range of carbon compounds. Batchelor tested materials by using a special apparatus in which the resistance of small samples of carbon could be measured as they were placed under varying amounts of pressure (Fig. 75). From May through July 1877, Batchelor tested hundreds of carbon compounds and mixtures, looking for one with the right electrical characteristics.

This draghunt for the right material may seem wasteful and even foolish, but one must remember that no one had developed a chemical theory that Edison could use to identify a form of carbon possessing the electrical properties he wanted. Consequently, such a draghunt was the only way Edison could secure the material he needed for the telephone. Furthermore, this search was hardly a hunt for a needle in the haystack, with Edison randomly trying everything and anything in the hope of finding the right material. Rather, thanks to the accumulation of insights from experimenting with the building blocks, Edison had a very clear definition of the needle he desired (i.e., the electrical characteristics of the material needed) and he had also

Figure 75. *Edison and Charles Batchelor used this apparatus to test different carbon materials in the spring and summer of 1877. The carbon sample was placed in the small cup in the center of the base. The arm from the right was then placed over the cup and different weights were placed on the arm. To measure the resistance, a current was passed through the sample (via the terminals at each end of the base) and then measured using a galvanometer.*

identified the haystack in which his assistants should look (i.e., carbon compounds).

Batchelor eventually found the ideal material in the soot deposited on the glass mantle of a kerosene lamp. This lampblack carbon possessed a resistance that could be varied under pressure from a fraction of an ohm to 300 ohms. Edison and Batchelor fashioned the lampblack into small carbon buttons which were placed directly underneath the needle connected to the speaking membrane. Finding both the volume and articulation good in this transmitter, Edison quickly set up a shed with dozens of kerosene lamps to produce the carbon and taught his men how to roll the lampblack into small buttons (Fig. 76).

Initially, Edison connected these carbon buttons to the diaphragm using a small piece of rubber tubing (Fig. 77). Based on his mental model of variable resistance, Edison still assumed that the acoustical vibrations must be transferred by some means (like a needle or tube) to the high resistance medium (carbon). Edison tested this rubber/carbon button transmitter on Western Union telegraph lines in the fall of 1877 and got satisfactory results. Because the company was anxious to get telephones into service in order to show the newly formed American Bell Telephone Company that it meant business, Edison had his men build several of these transmitters.

Figure 76. *In the carbon shed at Menlo Park, the lampblack carbon was scraped from the inside of the glass mantles of smoking kerosene lamps.*

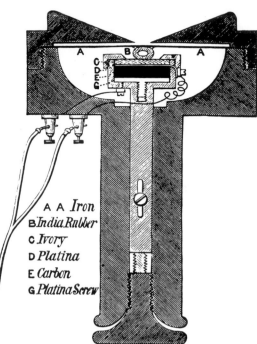

Figure 77. Developed by Edison in
the fall of 1878 the telephone
transmitter shown in this cutaway
used a carbon button and rubber
tube. The small rubber tube was a
substitute for the needles used in
earlier transmitters. Edison still
assumed that the vibrations of the
diaphragm had to be conducted to
the resistance medium, the carbon
button. The tube was first replaced by a small platinum spring
and then completely eliminated,
as shown in Figure 78.

A A *Iron*
B *India Rubber*
C *Ivory*
D *Platina*
E *Carbon*
G *Platina Screw*

However, Edison soon discovered that the rubber tubing lost its shape
and failed to conduct the vibrations. In response, Edison replaced the rubber tubing with a platinum spring, but found that even a delicate spring added
an extra musical tone to the signal. Next he tried thicker springs which gave
better results. This line of experiments led Edison to take the spring out
altogether and fasten the carbon button directly to a thick iron diaphragm.
This new arrangement gave superior results in tests conducted in early 1878
between New York and Philadelphia. It was capable of transmitting even
whispers loudly and distinctly, and it became the standard configuration of
Edison's carbon telephone transmitter (Fig. 78). Thus the course of these
experiments with building blocks led Edison to revise his mental model.
Previously, he had assumed that one could only get the phenomenon of
variable resistance by moving a needle through a high resistance material,
but these experiments showed him that variable resistance could be secured
by directly vibrating the resistance medium.

To enhance the operation of the carbon button transmitter, Edison placed
it in a more complex electrical circuit than Bell was using. In particular,
Edison included an induction coil in his telephone which functioned like
a modern transformer (Fig. 80). Edison connected his transmitter to the
primary side of this induction coil, and stepped up the signal current which
permitted the signal to be transmitted over longer distances. The stronger
current also insured that the electrical transmission could be heard more
loudly and clearly at the receiving end. The induction coil was a new
building block which Edison added to his telephone sometime in 1877, and
it became a standard feature of his telephone.

By the spring of 1878, both Edison and Western Union were satisfied
with the carbon telephone, and the telegraph giant organized the American
Speaking Telephone Company to install this telephone in several major cities.

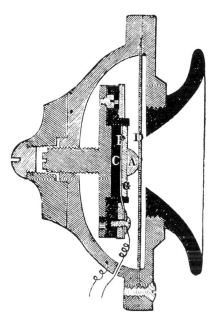

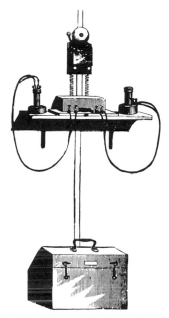

Figure 78. In this final version of Edison's carbon telephone transmitter, the platinum foil **P** serves as one electric contact to the carbon button **C**. The other electrical contact is the screw that rests against the carbon button.

Figure 79. In this Edison telephone from 1878, a magneto receiver (left) and a carbon transmitter (right) are located on the shelf. The box at the bottom holds the battery.

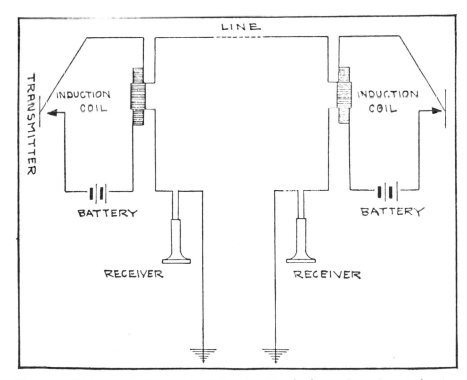

Figure 80. This circuit diagram for Edison's 1878 telephone shows how induction coils were used to boost the signal.

Edison assigned his telephone patents to Western Union for $100,000. Armed with Edison's patents, Western Union decided to take on the American Bell Telephone Company and eliminate this rival through patent litigation. However, this litigation proved to be long and difficult since the courts upheld Bell's basic telephone patent, and the Bell interests had secured a patent for a carbon transmitter from Emile Berliner. After a year of courtroom sparring, Western Union and American Bell reached an agreement. Western Union agreed to withdraw from the telephone field and to sell its existing telephone exchanges to American Bell; in return Bell agreed to pay Western Union a royalty of 20 percent for 17 years on the telephone rentals on their former exchanges. This agreement has often been viewed as a defeat for Western Union, since the company lost control of a new technology. Yet, as far as the telegraph giant was concerned, it was probably a victory; Western Union had forced Bell to pay it a sizeable royalty on an unproven technology. The company would enjoy a steady income and let Bell assume all of the risks of further perfecting the telephone. Western Union was able to secure these favorable terms thanks to the patents supplied by Edison.

Beyond the Carbon Transmitter: Spinoff Inventions in 1878
Although the development of the carbon transmitter was the predominant line of investigation in his telephone project during 1877, Edison also experimented with a wide range of transmitters. Some of these were modifications of his carbon transmitter, while others varied the resistance by using switches, capacitors, or batteries. Underlying all of these designs was Edison's mental model of variable resistance; thus, the designs reveal Edison's ability to generate a variety of alternative representations of his central idea. In terms of business strategy, Edison probably built and publicized these alternative transmitters in order to give Western Union broad coverage of telephone technology and to prevent Bell from being able to use any of these alternatives.

Yet these telephone inventions failed to exhaust the potency either of variable resistance as a mental model or of carbon and diaphragms as building blocks. Throughout 1878, Edison used this model and these building blocks to create a remarkable series of spinoff inventions. Drawing on the idea of the variation of carbon's resistance under pressure, Edison perfected first the carbon microphone and then the tasimeter for measuring very slight changes in temperature or motion. But even more exciting, Edison used the knowledge he acquired while working with the vibrating diaphragm of his telephone to invent the megaphone, a stethoscopic microphone, the phonomotor, and of course, the phonograph. With these many inventions we see how Edison could invent by manipulating the mental model as well as by building upon intimacy with the building blocks.

As fascinating as these many inventions were, Edison dropped them all in the fall of 1878 to take up incandescent electric lighting. With a new group of backers and substantial venture capital, Edison devoted much of his energy to this new endeavor. But during the winter of 1878–79, he returned to the telephone for one last fling. In England, Edison had permitted several businessmen to establish the Edison Telephone Company, which competed directly with the British Bell Company. While the Edison company held its own for a time, British Bell gained the upper hand when it received a broad British patent covering the principle underlying the magneto receiver. Since the Edison company was using a magneto receiver, it became necessary for Edison to develop an alternate receiver quickly. Just as he had done previously with the carbon telephone, Edison drew on an existing invention for a building block. In this case, he went back to his 1874 electromotograph, which was a sensitive nonmagnetic relay. This device worked on the principle that a changing electrical current varied the friction on an electrode as it rubbed against a moist chalk surface (Fig. 81). Edison adapted this idea to the telephone and created a receiver in which the listener used a small crank to turn a chalk drum in the receiver. As the drum rotated, the incoming electrical signal caused the rubbing electrode to vibrate, which in turn activated the diaphragm in the earpiece (Fig. 82). Although this receiver worked well, the Edison Telephone Company eventually merged with the Bell interests in England and the new company resumed use of the simpler magneto receiver. Nevertheless, the chalk drum receiver illustrates well Edison's ability to circumvent patent problems by substituting building blocks within a mental model.

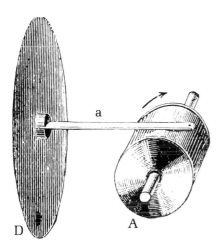

Figure 81. *This simple sketch shows the principle behind Edison's chalk drum telephone receiver. The chalk drum **A** is moistened with a solution of potash and mercuric acetate. The drum is connected to one terminal of a battery (not shown) while the electrode **a** is connected to the other terminal. As the drum rotates and current passes through it, the friction between the drum's surface and the spring electrode **a** varies proportionately with the amount of current in the circuit. If the current is fluctuating (as in a telephone circuit), then the electrode will vibrate rapidly and will produce sounds at diaphragm **D**.*

Figure 82. *A user of Edison's chalk drum telephone receiver from 1878–79 turned a crank (not visible) which rotated the chalk drum. The vibrating diaphragm is seen in the back of the center of the box.*

Conclusion
The story of Edison and the telephone shows that the act of invention may be viewed as the mingling of an abstract idea with tangible objects, of a mental model with building blocks. As in other creative endeavors, an inventor succeeds if he or she is able to manifest his or her mental model in terms of building blocks; it is very much like the artist who takes a vision in the mind's eye and uses color and shape to express it. For the telephone, Edison's mental model was the principle of variable resistance, and his building blocks were devices borrowed from other inventions such as the liquid rheostat, carbon rheostat, induction coil, and chalk drum.

One might expect that in the course of invention Edison would have moved from theory to practice, from mental model to building blocks. However, as this case shows, Edison obtained new ideas for his telephone by working closely with his building blocks. To be sure, he had a mental model of the telephone, but he chose to hold the model constant, introducing variations by manipulating the blocks. Thus, rather than playing with ideas and theories from the "top down," Edison shaped the telephone by working from the "bottom up." Such an approach may seem surprising to those of us conditioned by popular notions of science-based innovation; we have been led to expect that the scientist or inventor creates by applying scientific theory to a technical problem.

Because many tend to see the scientifically based "top down" approach as being rational, orderly, and efficient, it is easy to conclude that a "bottom up" approach must be the opposite—that it is irrational, disorderly, and inefficient. And rather than being based on careful thought, a "bottom up" approach may be viewed as based on luck or the mystery of genius. In Edison's case, this line of thinking has led people to view the great inventor's draghunts as the essence of lucky breaks. How could Edison be so crazy as to think he could find the right material for his telephone transmitter by simply trying dozens of different materials? But this approach was less of a gamble than might appear for, as this case has shown, Edison used the accumulation of insight from experiments with selected building blocks to set research parameters. His "hands on" familiarity with the carbon rheostat helped him determine what to look for when he had his assistants conduct a broad search for the right type of carbon. Lacking a theory for understanding the electrical properties of materials, how else could one identify the right material for an electrical application? Such "wild" searches are, indeed, very much part of contemporary research and development; for example, hundreds of scientists spent the better part of 1987 racing to "cook up" a ceramic material with superconducting properties.

The development of the carbon telephone helped Edison establish a method which he used with subsequent inventions at Menlo Park. In particular, Edison used several tactics from the telephone project in the course of inventing his system of incandescent lighting. For instance, he began the electric lighting project in 1878 by borrowing various electrical circuits from his telegraph devices, which he used to control the current in the lamp filament. Likewise, it is well known that upon discovering in October 1879 that carbon made a good, high-resistance filament, Edison mounted a major draghunt to again find the right carbon compound. Finally, it appears that Edison shaped his conception of an electric lighting system as he went along;

although he had a general mental model of the system at the outset, the details of the system only became clear as he and his staff worked closely with the lamp, generator, and other components. Here again, we see Edison working from the "bottom up."

All in all, Edison's work on the telephone at Menlo Park reveals new facets to his creative genius. But these facets, which show him as a manipulator of building blocks and a leader of well-planned draghunts, do not mesh nicely with the long-standing myth of luck and persistence. Nevertheless, it is only by looking at Edison in this multifaceted way that we can fully appreciate his abilities and at the same time learn what can be done to improve innovation in our own time.

Bibliography

This essay is based primarily on an examination of the telephone artifacts in the reconstructed Menlo Park Laboratory at Greenfield Village. To place these artifacts in context, we consulted the Edison manuscript collections in both the archives and library of Henry Ford Museum & Greenfield Village and the Edison National Historic Site in West Orange. The latter collection is now partly available on microfilm; see Thomas E. Jeffrey, et al., eds., *The Thomas A. Edison Papers: A Selective Microfilm Edition, Part I (1850–1878)* (Frederick, Md.: University Publications of America, 1985). Especially useful are the evidence and exhibits from Edison's telephone interference litigation on reel 11.

For additional information about Edison's telephone work, see George B. Prescott, *Bell's Electric Speaking Telephone: Its Invention, Construction, Modification, and History* (New York: D. Appleton, 1884; reprinted Arno, 1972) as well as Francis Jehl, *Menlo Park Reminiscences*, 3 vols. (Dearborn, Mich.: The Edison Institute, 1934–41) 1:100–155. Especially valuable in understanding some of Edison's telegraph experiments is Reese V. Jenkins and Keith A. Nier, "A Record for Invention: Thomas Edison and His Papers," *IEEE Transactions on Education* E–27 (November 1984):191–196.

Should the reader wish to learn more about the early history of the telephone, consult the following. Robert Bruce, *Bell: Alexander Graham Bell and the Conquest of Solitude* (Boston: Little, Brown, 1973); Bernard S. Finn, "Alexander Graham Bell's Experiments with the Variable-Resistance Transmitter," *Smithsonian Journal of History* 1 (1966):1–16; David A. Hounshell, "Elisha Gray and the Telephone: On the Disadvantages of Being an Expert," *Technology and Culture* 16 (April 1975):133–161; F. L. Rhodes, *Beginnings of Telephony* (New York: Harper, 1929); and A. R. von Urbanitzky, *Electricity in the Service of Man*, ed. R. Wormell (London: Cassell, 1886).

For those desiring to know more about the conceptual framework underlying this essay, we would recommend our own "The Cognitive Style of Inventors: Thomas Edison, Alexander Graham Bell, and the Telephone," Harvard Business School Working Paper No. 89-026. Also helpful is Thomas P. Hughes, "Edison's Method," in *Technology at the Turning Point*, ed. W. B. Pickett (San Francisco: San Francisco Press, 1977):5–22.

We wish to thank William S. Pretzer for his assistance in our study and photographing of the Edison artifacts discussed in this essay. We are also grateful to Robert Bruce and Keith A. Nier for clarifying several key technical points.

Figure 83. This photograph, which shows Edison (seated), investor Uriah Painter (left), and Charles Batchelor with a tin foil phonograph, was taken in 1878.

Drawing as a Means to Inventing: Edison and the Invention of the Phonograph

Edward Jay Pershey

TRYING TO UNDERSTAND how Thomas A. Edison thought as he invented his way into the annals of American history is indeed a herculean task. His patents, numbering over a thousand, cover a wide array of technologies. His productive career ended over 60 years ago, and his initial work on electric lighting, telephone, and phonograph occurred over 100 years ago. While many of the technologies we use daily have deep historical roots in Edison's work, the actual forms of the technologies—direct current lighting, mechanical sound recording, and electro-mechanical telegraphy and telephony, to cite a few—for the most part have changed dramatically. Yet Edison's reputation as a genius, a "Wizard," not only persists, but actually grows stronger as our culture becomes further separated from his. Fortunately, he left us ample information and data with which to attempt an interpretation of his way of inventing and the character of his creative thoughts. He left us thousands of drawn pages of laboratory notes.

Studies of creative thought have repeatedly pointed to the importance of what is called visual thinking. Interpretations of cognitive processes in the human brain no longer make simple splits between left and right brain, between language and art. More often highly creative thought is perceived as a combination of various "modalities" of thought. The role of visual thinking and techniques of graphic analysis in technical creativity cannot be minimized. One of the ways that historians have come to differentiate technology from science is to see that the end product of technology is something—some kind of artifact—that must work in a certain way in the day-to-day world. In pursuing such technological investigation, drawing the spatial relationships of artifact to artifact, and the arrangement of the parts of a single artifact, serves as a critical mode of thought.

Visual thinking is particularly important in the invention of new products. Drawing, more so than verbal skills, can effectively express the ideas of someone who is building a structure or machine and wishes to convey those ideas to another person. Technology does indeed have its own array of verbal jargon, as does any highly specialized profession. Much of this jargon is familiar to us—just think of the names of some of the parts of your automobile. But think about how often you have read the name of a part and then seen it. Your reaction may have been "oh, so that's what that is!" The visual identification may convey not only name, but function and spatial placement in the machine, something that the name of the part alone may not do.

For instance, simple sketches are useful to people trying their hand at household carpentry and repairs, or at assembling an item. Often, before we try to build or arrange something ourselves, we will make a little sketch to guide our work. Commercial assemble-it-yourself kits are less of a test of nerves if accompanied by well-drawn, clearly marked graphic instructions. Similarly, the better do-it-yourself manuals are largely graphic. Professional workmen summoned to tackle complicated projects for us may rely on more finished, detailed blueprints. This range of drawing in technology—from the simple sketch to blueprints—represents the various ways that the technological practitioner, amateur or professional, can manipulate the ideas behind the artifact before working in wood, metal, or whatever material. It is this manipulation of matter in space, without actual physical rearrangement of matter or space, that makes drawing so potent in technology.

Manipulation of the real world through the medium of drawing is a technique that was used at least as long ago as the Middle Ages, as is evident in the work and notebooks of architects who designed and constructed buildings and fortifications for the ruling classes. This technique continued to exert an important influence on technology, as these drawings were copied and circulated through the informal network of working architects. After the introduction of the printing press in the 15th century, many of these drawings were codified into compendiums of illustrations of machines, both real and imagined. This tradition of graphically solving engineering problems continued, and survives in the current techniques of computer-aided design and manufacturing. Not everything that is drawn is built, but everything that actually gets built is first *drawn*.

Edison, urged on by the lawyers who sought to protect him in patent lawsuits, began early in his career to set down his ideas both graphically and verbally. His first notebooks, dating from the early 1870s, contain carefully crafted, detailed sketches of the telegraphic apparatus that absorbed the energy of the young inventor. Other forms of Edison notebooks appeared as well. Notebooks containing more sketchy, cryptic entries were made ''on the fly,'' as records of ongoing work at the laboratory bench. These notebooks, placed conveniently around the lab, were always at hand to record the latest efforts. However, because these experimental books were not always used systematically, it was possible for different stages of the same experiment to be recorded in several different books and, conversely, any one book might contain snippets of several different experiments. Some entries were dated and signed, others not. Simpler scrapbooks with clippings from technical and popular journals, filled with images of the state of the art in specific areas of technology, were yet another form of Edison's graphic record.

Edison was not alone in keeping such notebooks. Daybooks—small, bound note pads—were commonly carried by all sorts of people at this time. In these small books individuals made assorted, often random, entries related to their lives: reminders, accounts, thoughts, sketches, and so on. Other inventors, contemporaries of Edison, also kept notebooks and made sketches. This practice became standard procedure in the first corporate research facilities that took shape at the turn of the century, with the notebooks becoming more formal and standardized. Similar notebooks are used regularly by modern corporate inventors and engineers; however their neat, standardized formats are unlike the relatively untamed Edison books. In fact, many of

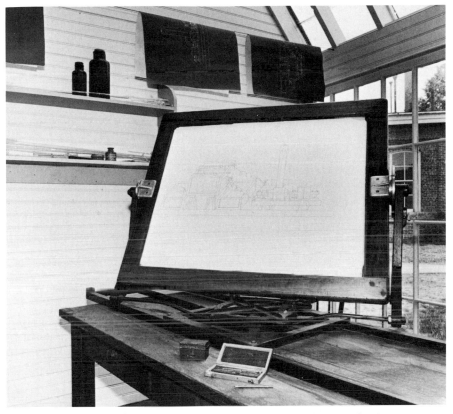

Figure 84. *Edison used some version of a cyanotype machine to duplicate technical drawings.*

Figures 85 and 86. *Draftsman Samuel Mott produced drawings for patents on a drafting table like the one in the reconstructed Menlo Park office building.*

Figure 87. *Edison strikes a confident pose in his Menlo Park Laboratory. Two tin foil phonographs of the kind discussed in this article are displayed in front of him. This engraving, based on a photograph taken at the lab, was published in the* New York *tabloid* The Daily Graphic *on April 16, 1878.*

Edison's sketches and drawings survive as loose sheaves on the back of poor quality brown paper, or the reverse of a piece of used stationery. The wealth of information is astounding, fascinating, and sometimes frustratingly indecipherable—especially the sketches that lack any kind of verbal or associated visual clue as to what he was doing. Many others are undated, leaving the viewer to wonder just where in the seemingly endless parade of his work the drawing should appear. But at least we have these drawings, and they provide at times a rather complete record of Edison's work.

For inventors like Edison, the essence of whose creativity lay in the production of an artifact of some kind, drawing their ideas out was an especially powerful technique. The freedom it gave in conjuring new ideas in spatial arrangement allowed the inventor to ''build'' prototypes in his mind before, or as, those models took shape in metal, glass, plastic, or wood. As is true in Edison's case, the prototypes have rarely, if ever, survived for us to handle and inspect. We must therefore rely on the two-dimensional representations in the drawings to learn about the experimental originals.

Studying drawings related to invention can be a path toward an understanding both of a particular technology and of the way that technological change comes about. Let us turn now to some drawings made by Edison in the course of inventing one of his most popular creations—the phonograph. In so doing, we can begin to see how the inventor invented and also understand what it was that he really did invent (Fig. 87).

Edison's invention of the phonograph has been popularly perceived as the inspired work of one or two afternoons in the fall of 1877 at his famous Menlo Park Laboratory in New Jersey. He had moved out to rural New Jersey from his Newark shop the previous year intent on establishing what we would call an ''R&D Lab,'' a research and development facility closely tied to manu-facturing operations. The lab was primarily devoted to telegraphic work, building on Edison's expertise in that field and taking advantage of the market for telegraph equipment for business applications. Edison knew full well that such a laboratory needed to fix its sights on practical, saleable inventions, since he counted on a base of marketable goods to provide a steady flow of cash into the laboratory so that he and his assistants could tackle several projects simultaneously.

As it happened, from the middle of the summer of 1877 through at least June of 1878, Edison worked on the idea of an apparatus to record and play back the human voice. The first successful form of the machine squawked out a few words in his lab probably during the last week of November 1877. The public unveiling and demonstration of the device at the offices of *Scientific American* in New York City occurred a week or so later. With that singular achievement, Edison acquired the title of the ''Wizard of Menlo Park.'' This wizard, though, did not have secret superhuman powers. He did, however, draw a lot.

The drawings related to the invention of the phonograph provide a special opportunity to observe the creative process of invention. The great majority of the Edison drawings from this phonograph work and his whole later career survive in the Edison Archives at West Orange, New Jersey, at the Edison National Historic Site. Here the National Park Service preserves Edison's last and largest laboratory, which he built in 1887 and where he worked until his death in 1931. In these archives, the drawings are preserved in context, together with the business correspondence, accounting records, photographs, and personal memorabilia contemporary with the drawings.

Some of the drawings are often ''annotated'' with verbal descriptions of what the inventor intended. Others are starkly devoid of any verbal clues. Some were quite clearly made in order to direct the construction of a machine; others were drawn from already fabricated artifacts; still others represented machines that were never built at all (Fig. 88).

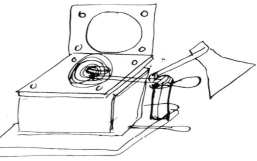

Figure 88. *This sketch, though undated, was most likely made in January 1878. One of two drawings on a single sheet, it shows a machine using discs of waxed paper or metal foil and a recorder-playback mechanism mounted on a swing-arm. This is from one of the few drawings in the series relating to the invention of the phonograph that has an explicit inscription from Edison that he wanted a model made just like the illustration and that he also wanted a drawing made from an existing machine.*

It is important to note that a number of the drawings were not made by Edison himself. The power of the Menlo Park Laboratory lay in the team of creative individuals that Edison had gathered there to work with him in a collaborative atmosphere. Some of the men were chosen because of their ability to make finished drawings, so that there may be different versions of a single drawing: a rather sketchy version by Edison, and a more finished piece by someone else for use in patent applications to the government. Then again, whole series of drawings in an associate's hand represent that person's work and ideas, rather than Edison's. Yet we know that throughout Edison's career, the experimental work at the lab was always directed by him. His ideas drove the creative work of his laboratories (Figs. 89, 90, and 91).

The importance of these drawings as historical evidence is amplified by the fact that the original artifacts no longer exist. Other than the "first phonograph," none of the intermediate experimental forms for the phonograph has yet been discovered in the collections of the various Edison museums or in the hands of private collectors. Only a few commercially made examples survive of the "tin foil" type of phonograph that was the end product of months of experimental work at the Menlo Park Laboratory during that summer and fall of 1877. While the artifactual evidence is scarce, the context for the drawings, as mentioned previously, is very rich. This is a powerful assemblage of striking images (Fig. 91).

The development of the phonograph derived from Edison's work on the telegraph and telephone. He was interested in ways of storing telegraph messages for later retrieval and playback—what we today would call information storage and retrieval. Information coming in or going out on a telegraph could be stored either as Morse Code or as alphabetical English. One form of this kind of device used rotating discs of paper on which a telegraphic relay would indent a spiral series of dots and dashes. This record of indentations could be read in turn by another mechanical relay mechanism which opened and closed an electrical circuit, generating an electrical signal to send out over the telegraph line (Fig. 92).

In the late 1870s Edison was also under contract to Western Union to develop a successful telephonic system that would not infringe on the basic 1876 telephone patent of Alexander Graham Bell. The Edison carbon button

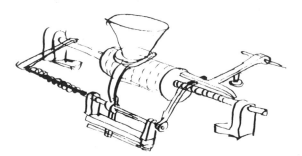

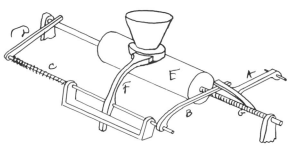

Figure 90. *This more finished version of Figure 89 was prepared by an assistant for the patent caveat notification. A caveat (the Latin word for warning) allowed an inventor to notify the U.S. Patent Office of the intent to file an application, thus establishing priority for work in progress.*

Figure 89. *Edison made a sketch of a phonograph with a dual feed-screw mechanism on February 3, 1878.*

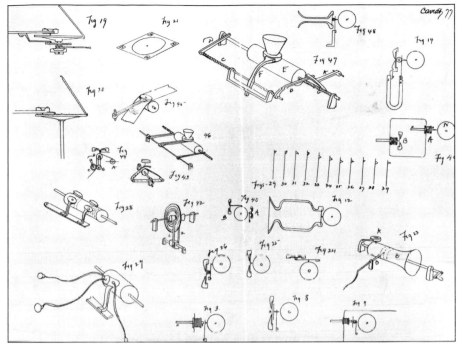

Figure 91. *This sheet of more carefully rendered drawings of the phonograph was prepared for a patent application in early 1878.*

transmitter proved to be a critical improvement on Bell's version of a transmitter, and when coupled to receiving mechanisms developed at the Bell laboratories, produced a system of telephony that worked well enough to be sold for general business use. The new telephone market grew out of the related market for private line telegraphs, which were used with printing devices to transmit gold, stock, and commodity prices, as well as regular

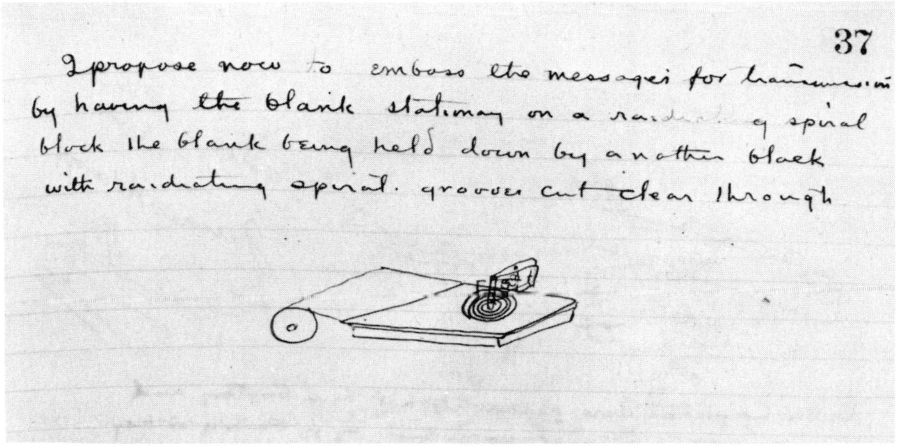

Figure 92. *In this illustration and caption (excerpted from a mid-1870s drawing) Edison described a spiral, disc-like device for recording Morse code telegraph messages. This kind of machine formed the basis for his later phonograph work.*

business communications. Storage and retrieval of telegraphed communications were important parts of those systems.

Edison was, therefore, intimately involved in devices that converted the mechanical vibrations of a telephone speaker into electrical impulses and vice versa. Edison worked from analogy with existing telegraph systems, an approach typical of him. Since the telephone itself was viewed, and often labeled, as a "speaking telegraph," it made sense to Edison to consider a record-and-playback device for the telephone similar to the one used in telegraphy. Edison was essentially looking to invent a phone answering machine. That goal proved elusive throughout his career, although the Edison company in the early 20th century made and sold a dictating machine designed specifically to record telephone conversations. In 1877, however, Edison discovered that the apparatus with which he and his staff had been struggling promised even more interesting capabilities. The new recording process they began to investigate in the summer of 1877 had little, if anything, to do with telephones.

About mid-July 1877 it became apparent to Edison that the human voice, as modified and transmitted on the "speaking telegraph" (telephone), ought to be able to be directly encoded, stored, and retrieved using modified telegraph equipment. The technique was analogous to that used in printing and repeating telegraphs, but entirely independent of the electrical telegraph or telephone. More importantly, the process of directly recording the sound entailed much simpler mechanisms than either the telegraph or the telephone, and was totally mechanical rather than electro-mechanical. That is, telegraph apparatus depended upon the conversion of electrical pulses into the mechanical movement of receiving devices. Likewise, information stored as bumps or perforations in paper needed to be reconverted into electrical signals to be sent back out on the telegraph line. At first Edison and his staff thought that sounds, like the human voice, might be captured, stored, and retrieved using similar conversion: sound to mechanical, to electrical, back to mechanical, and then reconstruction of the sound. However, experiments performed in the summer of 1877 indicated that merely the mechanical portions of this process would suffice to record and play back sound. In fact, dispensing with the electrical circuits made the whole business far simpler (Figs. 93 and 94).

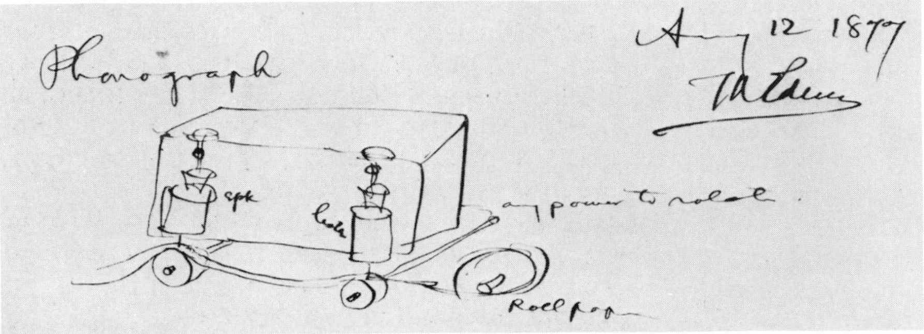

Figure 93. *One of the first drawings, August 12, 1877, clearly labeled "Phonograph," is little more than a ticker-tape printing telegraph device, modified with a telephone mouthpiece and speaker. Here Edison was only interested in the relationship of the sound diaphragm to the tape, and not in the motive power for driving the tape through the machine.*

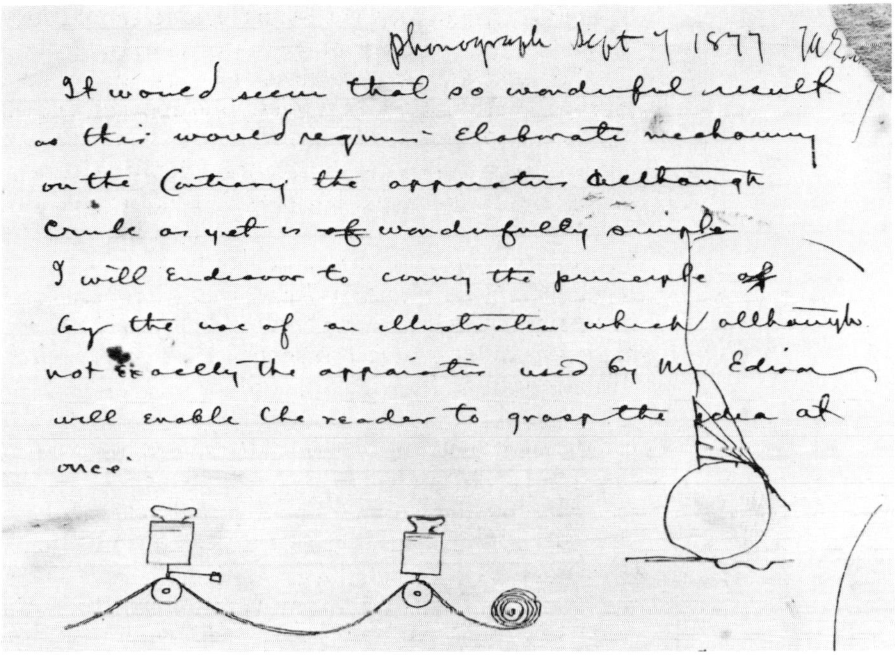

Figure 94. *Sometimes the drawings contain extensive written information. Here an assistant wonders at the simplicity of the phonographic effect in a passage apparently intended as part of a piece that would explain the work to the general public. Edison initialed the page near the date, September 7, 1877.*

Drawings made in August and September of 1877 (only weeks after Edison decided to investigate direct sound recording) show experimental mechanisms labeled ''phonograph,'' which consisted of various configurations of telegraphic equipment combined with telephonic diaphragms. Some of the drawings were probably of built apparatus, as indicated by the presence of assembly screws and nuts in the drawings. Other drawings were more clearly thought experiments, either with imaginary equipment or existing equipment used in new configurations. While Edison's style of drawing was not sophisticated, the sketches were energetic representations of his ideas as he constructed new machines in mental space, rotated these mental images to obtain various views of them, and put them into motion.

Other drawings delineated mechanical movement on a scale so small as to be microscopic. Edison needed to work out his experiments on paper, particularly when he was trying to understand the action of the stylus used to record the physical sound signal, since the tiny interface of stylus and recording surface defied direct observation. When looking at these series of small sketches, one can imagine Edison trying to envision just what kind of movement and effects a small needle might have and how that mechanical energy might be put to useful purpose.

One of the first drawings to be labeled with the term ''phonograph'' dates from August 1877. The number of drawings increased during the first week of September, with several dated September 7. The pace of work picked up through October and into November. By the middle of November 1877 the drawings indicate substantial work on the device. Although not specifically noted on any of the drawings, it can be inferred from the various configurations of the device that by mid-November it had begun to squeak out bits and pieces of prerecorded sound.

Popular histories of Edison's invention of the phonograph tend to portray his associates as naive accomplices who did not fathom what the "Old Man" was doing with the device. A classic example is the 1940 Hollywood film *Edison, The Man*. Starring Spencer Tracy as Edison, it showed him confounding his workers with the mysterious device and winning cigars from them in bets on its performance. The drawings, however, almost all of which were co-signed by one or more of his assistants, clearly show that not only did Edison know what effect the device would produce, but that his able associates also knew full well what a "phonograph" was supposed to do. Without their informed help, the great inventor could have done little more *than* draw.

It was during the third week of November 1877 that a final prototype machine was sketched out by Edison, and fashioned in the machine shop by John Kruesi, his chief machinist. Interestingly, one of history's classic "fakes" surrounds this event. A drawing in the Edison Archives in West Orange is dated August 12, 1877, and shows a rather complete machine looking very much like the final prototype of November. Scrawled across the drawing in Edison's firm hand is the directive: "Kruesi—make this—Edison." For many years this was touted, and published in books, as the drawing of the first phonograph. Oral history recordings made in the 1940s on a later T. A. Edison, Inc., "Ediphone" dictating machine tell a different story. The drawing, while actually dating from 1877 or 1878, was pulled out of a desk drawer in 1927 to be used in ceremonies surrounding the 50th anniversary of the phonograph. It was at that time, in 1927, that Edison added the inscription so that the drawing could be used in an advertisement for Edison disc phonographs. This original "fake" deluded Edison enthusiasts and historians for years!

The first successful prototype phonograph, also in the collections at West Orange, was actually a much simpler device than the one shown in any of the earlier drawn proposals. It looks unlike anything that we might take to be a sound recording machine. Totally the product of a machine shop environment, it even used metal foil for its recording surface. The sketch, dated

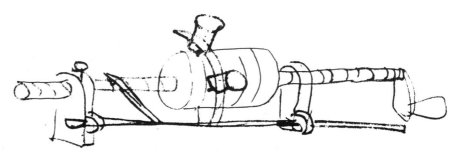

Figure 95. *Dated November 29, 1877, this drawing is the one that Edison most likely handed to his machinist John Kruesi, who fashioned the first phonograph in the machine shop at Edison's Menlo Park Laboratory. The cylindrical form of the machine, with its hand crank, feed screw, and dual recorder-speaker assembly, became the "standard" configuration for tin foil phonographs.*

Figure 96. The *"first phonograph"* is now in the museum collection of the *Edison National Historic Site in West Orange, New Jersey.*

November 29, may have been used to direct the manufacture of the device. It would seem logical that there would have been more finished mechanical drawings as well. In existence are several drawings that were made for the application for a British patent, but there are no others. Perhaps Kruesi had indeed worked from Edison's simpler sketch (Fig. 95).

The first phonograph was completely mechanical, using the vibration of the air from spoken words to move a steel needle stylus. This stylus was attached to a thin, steel diaphragm placed at the bottom of a short funnel-like mouthpiece. As the operator spoke into it, this diaphragm would vibrate in sympathy with the sound. The recording surface was lead or tin foil, overlaid on a brass cylinder that had a permanent spiral groove. The brass cylinder was mounted on a feed screw—with the pitch of the feed screw matching that of the groove—which was turned by hand. As the operator cranked the machine, the cylinder advanced steadily under the diaphragm-stylus assembly. The stylus, affixed to the center of the diaphragm, vibrated to the sound impinging on the diaphragm and deformed the surface of the foil to create a crude series of indentations which formed an analog of the sound hitting the diaphragm over time (Fig. 96).

For playback a reproducing stylus was run back over this series of indentations. This vibrated the stylus and therefore the diaphragm to which it was attached. The vibrating diaphragm disturbed the air to reproduce the sound. About 10 seconds of recorded sound was possible and the sound could be played back no more than two or three times before the stylus wore the deformations in the foil into an undifferentiated groove. Since the device was hand-cranked, the operator had to turn the cylinder at a steady rate while simultaneously yelling into the mouthpiece.

The resulting recordings, shallow indentations on the metal foil, could not be removed from the machine without being destroyed. A few examples of "tin foil" records from 1878 survive, but none has been made playable. The sonic record is mute (Fig. 97).

What the drawings of the device do not show was the rather complicated process involved in using it. This can only be discovered by actually using

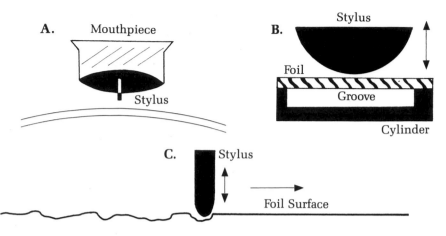

Figure 97. **A**: *The mouthpiece is a simple funnel into which the operator speaks. The sound hits the diaphragm, moving the stylus along a surface of thin metal (tin) foil which is overlaid on a solid metal (brass) cylinder.* **B**: *An enlarged section of the foil-cylinder arrangement shows the open groove in the cylinder under the foil. The stylus is made to move along foil just above the groove.* **C**: *The stylus moves up and down in response to the movement of the diaphragm, pushing down on the foil into the groove. The indentations in the foil form a series of "bumps," the analog of the sound hitting the diaphragm over time: a sound signal.*

the machine, which is not a mode of investigation normally permitted for a museum artifact of this importance. Luckily, in 1927 replicas of the original were made at the Edison laboratory in West Orange, under Edison's guidance. One of the several replicas fabricated is maintained in working order at the museum in West Orange.

Coaxing a recording out of this device is not easy. Although the process is, in essence, simple, it takes dexterity and a fair amount of luck to achieve success. Wrapping the foil around the cylinder tightly enough to produce a taut recording surface without tearing the foil is a tedious, and not always successful, operation that takes several minutes. Yelling into the machine and cranking it evenly has its amusing aspects. To put it mildly, the machine can be finicky. The quality of recording depends on so many variables as to defy analysis from attempt to attempt. Descriptions of the original demonstrations confirm that things were no easier in 1878.

The cylindrical form of the machine proved to be the easiest to design and make operable. It was, however, only one of three possible forms. From the beginning Edison had focused his attention on two more likely candidates based on designs already used in telegraph systems: discs and continuous tape. Even after the success of the cylinder machine, Edison continued to develop these other formats. The disc offered the possibility of being able to remove the recordings and save them. On the other hand, it posed problems of maintaining a constant speed for the stylus as it moved across the spinning disc from the outside inward. The tape format, while solving the problem of constant stylus speed, produced a recording that was fragile and deformed almost immediately as it was produced. By February or March of 1878 only the cylinder configuration remained (Fig. 98).

The new phonograph of 1878 was barely more than a toy. When demonstrated to the public, however, it caused great excitement for it did what no other machine had ever done: it recorded, stored, and played back the human voice! Edison set up manufacturing of the new phonograph, and licensed its manufacture to a few companies. A marketing and development company was set up to improve the device. Ways of driving it by some sort of motor and increasing the recording time were investigated. Different kinds of diaphragms and styli were tried, with thin mica discs proving best for the former, and steel needles for the latter (Fig. 99).

Several versions of this device were built, most with longer cylinders than the original, which permitted correspondingly longer recordings. These demonstration models, which also reproduced with louder volume than the original, were used on the lecture circuit in 1878 to astound people for the price of admission. Some smaller "parlor" models were also made and sold, but these had all the limitations of the original invention.

The inherent limitations of the device restricted it to the role of a fascinating novelty. By the end of 1878 Edison had turned to more interesting and promising investigations (electric incandescent lighting), and the Edison Speaking Phonograph Company was moribund. Even the original machine was casually lent out for a demonstration, and ended up in The Science Museum, London, England. In the late 1920s Edison petitioned the museum through the United States government to get his phonograph back. (The rep-

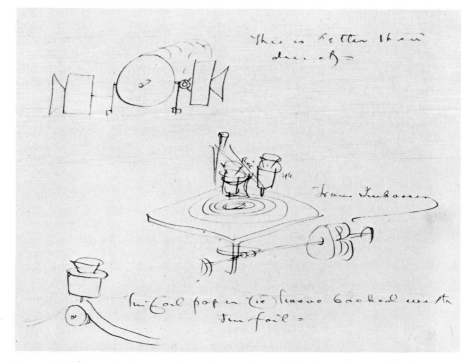

Figure 98. *These sketches, made on December 3, 1877, show three viable forms for the phonograph: cylinder, disc, and tape. (Only the cylinder format had been successfully demonstrated.) All forms used a metal (tin) foil as the recording medium.*

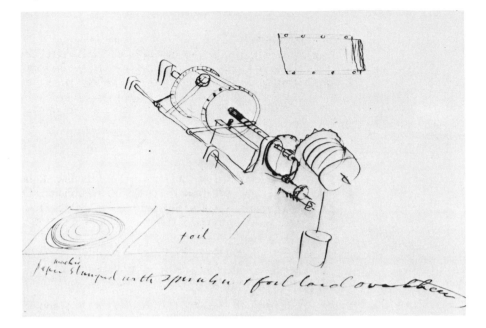

Figure 99. *This is one careful drawing of a weight-driven phonograph with tractor feed pins, which guided longer sheets of tin foil over the cylindrical mandrel. Next to this device is an illustration of a disc format tin foil record. These drawings were made in January 1878.*

licas were produced on its return to West Orange.) Successful phonographs, based on the same principle, but using different configurations and materials, were developed in the 1880s by researchers at Bell's laboratory and at Edison's new West Orange facility.

The later "reinvention" of the phonograph in the 1880s and 1890s also involved sketches, drawings, and blueprints. Edison's initial work in 1886 was entered into a notebook which he kept at his Florida winter home. This book contained more than the phonograph work, including, as well, various sketches of electrical systems, his ideas for a new research laboratory, devices to aid the deaf, doodlings of wicker furniture, and even the solar system portrayed as a giant electric dynamo.

Edison was on a honeymoon with his new bride, Mina Miller Edison, and his creative energy was reviving after the tumultuous and exhausting early years of establishing electric light and power stations across the country. His mind was racing with new ideas, all of which took the form of visual art in his notebook. Their graphic energy makes these Edison drawings engaging in themselves and gives them an artistic dimension that goes beyond their initial purpose as aids to invention.

For those who would find the subject of technological innovation forbidding and complicated, the Edison sketches offer a way to understand how at least one kind of technological invention occurred. The creative process of invention shows itself to be not the sleight of hand of a powerful wizard but, rather, an impressive combination of technical skill, knowledge of the material world, and an ability to work out ideas in graphic terms. The Edison drawings are, as it were, windows into his mind through which we can begin to appreciate the work of one of the world's foremost inventors.

Bibliography

There is very little published material about Edison and invention of the phonograph, other than the biographies and other general books about Edison. My information for this article derives largely from those works and from the original drawings and notebooks, some of which I have shared with you in this short article. The Edison Archives in West Orange, New Jersey, offer a great wealth of untapped materials for historians. The general public can visit the Edison National Historic Site where the National Park Service conducts tours and provides a carefully constructed interpretation of the site and Edison history in general. The original "first" phonograph is on display as part of the tour.

General histories of the phonograph are few. Roland Gelatt, *The Fabulous Phonograph, 1877–1977* (New York: Macmillan Publishing Company, 1977) is a readable, popular level survey. Oliver Read and Walter Welch, *From Tin Foil to Stereo: Evolution of the Phonograph* (Indianapolis: H. W. Sams, 1976) is a detailed history of the technology, but doesn't provide a very useful social or broad technological context. Daniel Boorstin, *The Americans: The Democratic Experience* (New York: Random House, 1973) has some useful social history of recorded sound, especially in the context of the development of the mass production of consumer goods and changing ideas about space and time.

The use of drawings in technology was the subject of an article by Eugene Ferguson, "The Mind's Eye: Nonverbal Thought in Technology," *Science* 197 (1977):827–836. The article is well written and enjoyable. Brooke Hindle's fine book, *Emulation and Invention* (New York: Norton, 1981), shows the importance of drawing and visual creativity in the work of Samuel Morse and Robert Fulton. Books on art—popular "coffee table" books—are easy to find, but these rarely include technical drawings. An unusual collection of reprints of famous blueprints is *Blueprints: Twenty-Six Extraordinary Structures* (New York: Simon & Schuster, 1981).

The use of drawings to express ideas has been the subject of many books, though few of them discuss inventors and engineers. One of the more popular and better known is Betty Edwards, *Drawing on the Right Side of the Brain* (Los Angeles: J.P. Tarcher, Inc., 1979). Howard Gardner, *Art, Mind, & Brain* (New York: Basic Books, 1982) suggests some interesting connections between adult creativity and children's drawings.

Vera John-Steiner, *Notebooks of the Mind: Explorations of Thinking* (New York: Perennial Library, Harper & Row, 1985) presents one of the best of the overviews for understanding how creative people use notebooks and sketches. She does not treat Edison specifically, however, since her work is derived from interviews with current inventors, artists, etc.

The task of interpreting Edison technical drawings involves more than the efforts of any one person. I want to mention the many people who are involved in Edison scholarship and who have informed not only my limited study of Edison, but also have directly influenced my views of technology and culture: Mary Bowling, formerly archivist at the Edison National Historic Site; Reese Jenkins, editor of the Thomas A. Edison Papers at Rutgers University; Paul Israel, also of the Edison Papers; Andre Millard of the University of Alabama, Birmingham; W. Bernard Carlson at the University of Virginia; and Thomas Jeffrey, editor of *Thomas A. Edison: A Selective Microfilm*. I especially want to thank Eric Olsen, formerly of the Edison site, who helped me prepare the illustrations from the Edison originals.

The Modernity of Menlo Park

David A. Hounshell

WHAT ARE WE TO MAKE of Edison's Menlo Park Laboratory? Was it, as some have contended, the first true industrial research and development laboratory in the United States? Or was it, as others have countered, something less, at best a proto-industrial R&D laboratory, at worst simply a greatly enlarged and well-furnished workshop of an eccentric though brilliant independent inventor? For today's visitor to Greenfield Village, a walk through the Menlo Park complex is apt to evoke a sense of distance from the present. The main laboratory's very textures—its painted walls; wooden floors, ceilings and furnishings; and its uneven levels of light—speak of a different era. Today's industrial R&D laboratories seem colder with their polished surfaces of glass, stainless steel, plastics, and enamels; their abundant fluorescent lighting; and their array of computers, cathode-ray tube monitors, and digital display instruments.

But textures can deceive; artifacts can miscommunicate unless their context—the history of which they are a part—is understood. The *historical* Menlo Park (as distinct from the *physical* Menlo Park reconstructed in Greenfield Village) provides a profound statement of modernity. The establishment, growth, operation, and demise of Menlo Park tell us a great deal about how technology today is developed and how new industries are created. Moreover, Menlo Park's history provides some striking parallels to the history of 20th-century corporate research and development laboratories. This essay seeks to explore the modernity of Menlo Park by interpreting its history in the context of its time and that of our own.

A New "City upon a Hill"?

When the Puritans set sail aboard the *Arbella* for Massachusetts Bay in 1630, they sought to establish a new community radically different from any that had ever existed, one that would serve as a model for the world. As that community's leader John Winthrop pointed out to this fellow believers, "wee shall be as a Citty upon a Hill, the eies of all the people are uppon us." The Puritans had, as the Reverend Samuel Danforth exhorted in a sermon delivered 40 years later, journeyed on an "Errand into the Wilderness." [Quoted in Perry Miller, "Errand into the Wilderness," in idem., *Errand into the Wilderness*, pp. 11, 2.]

In 1876, 29-year-old Thomas Alva Edison journeyed on his own errand into the wilderness when he abandoned his partnership in industrialized Newark, New Jersey, and established what he, or at least a reporter, called an "invention factory" in the rural hamlet of Menlo Park, New Jersey. Edison had pursued invention in both Boston and New York before he joined with William Unger in Newark in 1870 to develop telegraphic instruments for the

Figure 100. *Edison appeared haggard shortly before taking time off from Menlo Park in 1879 to travel with a group of prominent scientists on a scientific expedition to the American West.*

Western Union Telegraph Company and the New York Gold Exchange. But his designs for Menlo Park were far more ambitious. Edison vowed publicly to produce in his new laboratory "a minor invention every ten days and a big thing every six months or so." [Quoted in Matthew Josephson, *Edison*, pp. 133–34.] With such a pledge, there can be little doubt that Edison sought to make Menlo Park a "city upon a hill" for the industrial United States if not the world—a beacon pointing out the path of progress. By 1879 Edison's vision for Menlo Park had been realized; it was, as Edison's intense competitor Alexander Graham Bell noted that year, "a celebrated laboratory," perhaps the most famous laboratory in the Western world. [Alexander Graham Bell to Thomas A. Edison, May 25, 1879.]

Three years later, Menlo Park was all but dead, and by 1885 Edison had totally abandoned it. Had Edison failed? No, he had simply succeeded beyond his original vision; success, not failure, had undone Menlo Park. As was true of the later Puritans of the Massachusetts Bay Colony, Edison's Menlo Park laboratory of the early 1880s signaled something quite different to the world from what Edison had promised in 1876 when he journeyed to that rural New Jersey crossroads.

Why had Edison chosen such a remote locale for his laboratory and how did he proceed on his errand once he moved to Menlo Park? Edison's entire inventive career had been carried out in an urban setting. The American city was the principal locus of technological change during the last quarter of the 19th century; here is where the technological systems of the age were first developed on a large scale: telegraph, telephone, and electric light and power (all technologies to which Edison contributed). For whatever reasons, Edison sought distance from the city, possibly because urban rents were too high and city conditions did not allow Edison the kind of control over the entire inventive process that he wanted.

Here one cannot fail to observe the parallels between Edison's establishment of his laboratory in Menlo Park and the dozens of corporations that built research and development laboratories in rural areas in the 20th century, especially after World War II. The intent underlying the choice of such isolated locations for these corporate R&D facilities was that the research function of the corporation should be unfettered by either the bureaucracy of the corporate office or the day-to-day troubles of the manufacturing units; scientists could do better science if removed from worldly affairs.

But Edison's retreat from the city was by no means a retreat from the world, just as the research-driven corporations in fact never fully retreated from the world. Edison had hardly moved to Menlo Park before he let the world know about his ambitious goals for his new laboratory. But he sought more than "a minor invention every ten days and a big thing every six months or so." Edison wanted scientific prestige for his laboratory, not just patents and the monetary rewards they might offer. Without understanding this part of Edison's plan for his laboratory, one cannot understand the totality of the Menlo Park experience.

Seeking Scientific Recognition

Why scientific recognition? Initially, Edison hoped to right what he considered a wrong; he sought to reestablish among "scientific men" the credibility he had lost when he became embroiled in a controversy over the "etheric force." In November 1875, while conducting experiments in acoustic telegraphy, Edison observed bright sparks coming from part of his apparatus. After further observation and experimentation, he concluded, "This is simply wonderful, and a good proof that the cause of the spark is *a true unknown force*." [Edison Laboratory Notebook, December 12, 1875, as quoted in Josephson, *Edison*, p. 128.] Eager for instant fame, Edison informed the newspapers that he had found "a new force, as distinct from electricity as light or heat is." [*Scientific American* 34 (January 1, 1876), 17.]

What followed was an ugly battle of printed words not only about Edison's "discovery," but also about Edison himself and his method of announcing his presumed discovery. By the time he had gotten his Menlo Park Laboratory opened and running smoothly, Edison appeared to many to be just another charlatan, not unlike the wild-eyed inventors of the perpetual motion machines that were capturing the attention of many Americans. Edison's "etheric force" had been just as resoundingly disproved as the various perpetual motion machines.

Edison needed to gain legitimacy for himself and his laboratory in the American and European scientific communities for both psychological and practical reasons. He wanted to be viewed as legitimate so that he could secure scientists' help with his inventive work, obtain their expert testimony in the event of any patent disputes, and secure them as "public defenders" of his inventions. Satisfying this need was but another part of his errand in the wilderness.

In this respect the modernity of Menlo Park is quite striking. When, soon after the turn of the century, American corporations established formal research and development laboratories, their creators struggled—generally without great success—to recruit the highest caliber of scientists from academia. Academic scientists often viewed corporate research laboratories with contempt, dismissing them as places where "best" science could not be pursued because it was tainted by corporate objectives. For their part, corporations struggled to raise the visibility of their laboratories in the scientific community and to overcome the deep-seated prejudices of academics against industrial research. Even today, these prejudices remain, and the scientific work of such firms as General Electric, Du Pont, Kodak, AT&T, and IBM is constantly measured against the research done in universities.

Edison worked hard at building his reputation as a scientifically literate inventor during his days at Menlo Park. He followed closely the publications of American physical scientists and corresponded with those to whose work he believed he could contribute. For example, he developed a great interest in the work of Henry Draper, a pioneering American astrophysicist, and during 1877 corresponded with him frequently about Draper's discovery of oxygen on the sun. Draper invited Edison to tour his laboratory and at one point told the inventor, "I hope you will have time to print something about scints [scintillations?]; it is of value theoretically and practically." [Henry Draper to Thomas A. Edison, August 9, 1877.]

Edison's greatest scientific ally, however, was George F. Barker, a professor of physics at the University of Pennsylvania. Barker's was not a great scientific mind, but at least he closely followed the scientific and technical literature. More importantly, Barker was a member of the National Academy of Sciences, a group he regarded as the "highest scientific body in the country." [George Barker to Thomas A. Edison, November 3, 1874.] Extremely eager to have working models of any of Edison's latest inventions, Barker often praised him for his scientific acumen; he considered the inventor to be "extremely original and ingenious." [George Barker to Thomas A. Edison, September 30, 1877.] In 1880 Barker told a visitor that he admired Edison "not only for [his] inventive genius but for his scientific ability" and admitted that he himself was "not scientific enough to catch Edison trifling."

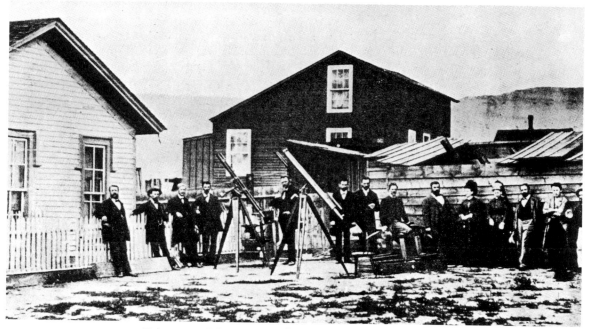

Figure 101. *Edison, second from the right, posed with other scientists and their scientific apparatus in Rawlins, Wyoming, while on their expedition to view the solar eclipse.*

[Charles H. Ames to Sarah Farmer, February 1, 1880, Moses Farmer Papers, Special Collections, University of California, Los Angeles.]

Indeed, Edison was comfortable communicating with scientists as diverse in their interests as Charles Darwin and William Thomson (Lord Kelvin), who regarded Edison as "one of the leading Electrical geniuses of the day." [J. C. Reiff to Thomas A. Edison, April 2, 1878.] Edison's invention of the phonograph especially strengthened his scientific reputation; as the author of the *Handbook of Electrical Diagrams* wrote him with great hyperbole, "it has made you foremost among the scientific men of the world [and] you will rank with Sir Isaac Newton and Galileo and LaPlace." [Charles H. Davis to Thomas A. Edison, March 31, 1878.]

Prestige but Also Profits

Edison recognized that even the best scientific reputation in the world would not put bread and butter on his table or, as one research director of the Du Pont Company used to say, "ring the cash register." [David A. Hounshell and John Kenly Smith, Jr., *Science and Corporate Strategy: Du Pont R&D, 1902–1980*, p. 150.] To be financially successful, Edison had to produce paying inventions. Edison's problem was one that all industrial research and development laboratories face: trying to balance the needs of doing good science with the absolute necessity of making the laboratory pay. Since the breakup of the Bell system, the management of AT&T Bell Laboratories has found itself precisely in this situation, and the result has been rapid change in the outlook and mission of that distinguished research enterprise. ["Deregulation Drags Bell Labs out of its Ivory Tower," *Business Week* (December 3, 1984) pp. 116, 121, 124.]

What impresses the reader of Edison's Menlo Park laboratory correspondence in the period from 1876 until the fall of 1878 is the large number of projects on which Edison and his staff were at work. The technologies of multiplex telegraphy (including the quadruplex to which Edison contributed significantly and the new harmonic types) and telephony were changing very rapidly, and Edison stayed on the cutting edge by following the work of other inventors and seeking alternative routes to the desired ends. Menlo Park's work in these areas led to Edison's invention of the carbon telephone transmitter, the priority for which was hotly debated with Englishman David Hughes. Edison's work on telephone transmitters led in turn to his first really big discovery, the phonograph, in 1877, and a period of intensive effort to perfect the invention. To enumerate the many other projects undertaken by Edison would be of marginal utility for our purposes, but the list is quite impressive and so are the results.

Moreover, Edison had to cope with myriad problems brought on by his very success as an inventor: patent applications, interference proceedings, publicly conducted debates about the priority to be accorded various inventions, and fame. This latter problem grew to gigantic proportions after Edison disclosed the invention of the phonograph. Anybody who had ever known or worked with Edison from his boyhood through his days in Newark, New Jersey, wrote to Edison to congratulate him and, mostly, to ask for a variety of favors. Some wanted money, some a job; others wanted him to speak at some gathering, or asked to borrow his instruments. With Edison publicly pegged as "the Wizard of Menlo Park," more and more people sought tours of his laboratory. For a long while, Edison obliged. At one point in mid-1878, Uriah Painter, Edison's agent for the phonograph, told the famous inventor, "you ought to close your doors at Menlo [Park] to all strangers and only show your hand when you are ready for the market and then its too late for thieves to get your things." [Uriah Painter to Thomas A. Edison, June 12, 1878.] Shortly thereafter, Edison issued a tongue-in-cheek call in the *New York Sun* for "musty tobacco for the use of malefactors, general depredators, and visiting scientists." [*New York Sun*, August 29, 1878; see also Lorillard & Co. Tobacco Works to Thomas A. Edison, August 29, 1878.]

The extraordinary, varied demands on Edison and the laboratory's failure to bring the much-hailed phonograph to a commercially viable state of development took their toll. Edison essentially "burned out." Seeing his condition, George Barker arranged for him to join a team of scientists, led by Henry Draper, who journeyed to the West to observe a solar eclipse in the summer of 1878. Edison had earlier corresponded with the astronomer Samuel P. Langley about developing fine measuring instruments for astronomical investigations. Langley had encouraged him, saying that "you would not perhaps produce anything commercially paying, but you certainly would confer a precious gift on science." [Samuel P. Langley to Thomas A. Edison, December 3, 1877.] Now Barker proposed that Edison employ his "tasimeter" (a highly sensitive heat-measuring instrument) to make measurements of the eclipse as part of the expedition's observations. In addition to interacting with the scientists on the trip, Barker proposed that Edison present a paper on his tasimeter trials at the August meeting of the American Association for the Advancement of Science (AAAS), which was to convene in St. Louis. Edison agreed and took an extended leave from his Menlo Park Laboratory.

The inventor's journey to the West provided him with more than needed rest and sustained interaction with some of America's best scientists. It gave him Menlo Park's next big project—the one that would enshrine Edison in the Pantheon of great American inventors. On the journey, Barker talked extensively with Edison about an emerging technology to which he thought Edison might contribute: electric lighting. Inventors in France, England, and the United States were trying to perfect an electric lighting system, but none had yet succeeded. Barker, who seems to have possessed an insatiable appetite for new technologies, had recently visited the workshop and factory of William Wallace in Ansonia, Connecticut, where the latter was developing his system. Wallace was a wire manufacturer, brass and copper founder, and commercial electroplater, and had teamed up with prominent electrical inventor Moses G. Farmer, who was one of three inventors claiming the "discovery" of the dynamo electric principle. Wallace and Farmer had begun to manufacture dynamos for electroplating applications, and both were trying to invent electric lighting systems to be fed by their dynamos. Excited about the prospects of contributing to the rapidly emerging electric lighting technology, Edison jumped at Barker's offer to take him to Ansonia, where Wallace would "show us all sorts of nice things." [George Barker to Thomas A. Edison, September 2, 1878.]

Because his wife became seriously ill while he was still in the West, Edison was forced to cut short his trip by passing up the AAAS meeting in St. Louis. The inventor returned to Menlo Park to face not only a wife suffering from "nervous prostration and fainting," but also all of the old problems like patent interference proceedings and bitter public disputes with rival inventors over their respective claims. [S. L. Griffin to Thomas A. Edison, August 5, 1878, and August 16, 1878.] He also had to worry about paying the bills, including the salaries of his small band of helpers. What motivated Edison now, however, was his desire to work on electric lighting.

Upon his return to Philadelphia, Barker arranged for Edison to tour Wallace's workshop on September 8, 1878. [Barker to Edison, September 2, 1878.] Barker and Charles F. Chandler, professor of chemistry at Columbia University, accompanied the inventor to Ansonia, where the three saw everything Wallace had done in electric lighting. Edison was impressed. As he told a reporter for the *New York Sun:*

> *I saw for the first time everything in practical operation. It was all before me. I saw [that] the thing had not gone so far but that I had a chance. I saw that what had been done had never been made practically useful.* [New York Sun, *October 20, 1878.]*

Edison, the inventor of telecommunications technologies, now turned his and Menlo Park's attention to electric lighting.

Focusing R&D
Edison's shift to electric lighting deserves emphasis. In the management of industrial research and development, the degree to which a laboratory can focus its research on an emerging technology often correlates with the success of that laboratory in contributing to that technology. (An outstanding example is the Pioneering Research Laboratory of the Du Pont Company's

Figure 102. *In the late 1880s, after being abandoned by Edison, the Menlo Park complex began to take on a ghostly aura.*

Textile Fibers Department in the period from 1940 to 1970. See Hounshell and Smith, *Science and Corporate Strategy*, pp. 384–444.) In early September 1878, Edison could have chosen to undertake research in electric lighting while continuing to work in telegraphy and telephony, the latter of which was a rapidly emerging technology. But Edison chose instead to focus attention almost exclusively on electric lighting. [Thomas A. Edison to George Gouraud, September 11, 1878.] As Edison himself realized, he was perhaps foregoing the surer gains to be made by successful work in telecommunications for longer-term, riskier but potentially greater payoffs in electric lighting.

Why focus his work on electric lighting? Edison made this decision only after he had, in his words, "struck a bonanza in Electric Light—indefinite subdivision of light." [Thomas A. Edison to Theodore Puskas, September 22, 1878.] While at Wallace's shop in Ansonia, Edison had correctly identified the subdivision of Wallace's 4,000-candlepower electric lights as the critical factor in making electric lighting commercially attractive. As Edison said, "The intense light had not been subdivided so that it could be brought into private houses." [*New York Sun*, October 20, 1878.] Within a week of his visit to Wallace's factory, Edison tumbled to what he believed was the solution to this technical problem. Had Edison not had a big "hit" in his research, it is unlikely that he would have made the boastful public statement about the success of his electric lighting system that appeared in the *New York Sun* on September 16:

> *Singularly enough I have obtained it [the electric light] through an entirely different process than that from which scientists have sought to secure it. They have all been working in the same groove and when it is known how I have accomplished my object, everyone will wonder why they never thought of it. . . . I can produce a thousand—aye, ten thousand—from one machine.*

Edison continued by describing how he would use gas light fixtures for his electric lights and how wires would run underground from a big generator into buildings and houses. Such statements, suggest the biographers of Edison's electric light, make clear that his "vision of a complete electric lighting system" was present from the outset. [Robert Friedel and Paul Israel, *Edison's Electric Light*, p. 14.]

Edison's pronouncement not only put on record his intent to deliver a commercially viable electric lighting system (in a few weeks, he said!), but it also provided him with the means to attract the necessary capital to carry out this venture. In this respect Edison's Menlo Park operation begins to look less and less like a corporate industrial research and development laboratory and more and more like many of the new ventures that are publicized in today's business journals.

When Edison first established the Menlo Park facility, his primary work was for the telegraph industry, especially the Western Union Telegraph Company. One could easily view the laboratory as an appendage to Western Union, a captive whose research and development work was focused on Western Union's short-term technical needs. But Edison's venture into the electric lighting field suggests that he was seeking to found what is today called a high technology start-up company, an enterprise resting on the pioneering technical ideas of an individual or small group of individuals and backed by venture capitalists. Indeed, this image is highly accurate, because the New York capitalists who sought a piece of Edison's bonanza did so through the establishment of the Edison Electric Light Company, which became the principal means of support for the Menlo Park Laboratory.

In exchange for half of his interest in all his electric lighting work, Edison received a $50,000 advance for research and development expenses. Moreover, he was to earn a royalty of $30,000 per year if the venture proved successful. Although the venture capitalists erected no explicit milestones for Edison to reach before they would provide him with additional funds, Edison was well aware that he had to produce results—to reach milestones—in order to continue to receive funding. As the historians Friedel and Israel have aptly noted, "For the next year the efforts at Menlo Park had to be directed not only to the creation of the practical light Edison wanted, but also to producing evidence that could convince investors of progress toward a profitable invention." [Friedel and Israel, *Edison's Electric Light*, p. 22.] Edison frequently had to renew his backers' faith in him and to rekindle their hopes for electric lighting by demonstrating progress toward his goals at the Menlo Park Laboratory. Certainly these backers expected Edison to devote himself and his laboratory exclusively to electric lighting. Edison's neck was really on the line, and this led research and development activities at the Menlo Park Laboratory to assume a very sharp focus.

With focus came thoroughness. Before they sank any more money into electric lighting, Edison's backers sought more information on "the state of the art." Their desire coupled neatly with Edison's own request to his patent attorney for information on all U.S. patents related to electric lighting and dynamos. [Thomas A. Edison to Lemuel W. Serrell, October 31, 1878.] In response to a request by Edison, the backers hired a young German-trained physicist named Francis R. Upton to conduct a thorough literature search on electric lighting. (Upon completing his search of patents, scientific papers and articles, and popular press items located in Boston and New York libraries, Upton joined Edison's staff at Menlo Park.) [Details on Upton's hiring are in Friedel and Israel, *Edison's Electric Light*, pp. 27, 29.]

Edison also initiated subscriptions to many technical journals that published work related to electric lighting. At the same time, he established formal procedures for record-keeping on all laboratory work; numbered laboratory notebooks would provide the laboratory with a complete record of its work, thereby facilitating that work and protecting it legally. These three steps—thorough literature searches, keeping up to date on the emerging science and technology, and careful documentation of work done in the laboratory—were among the central elements of the "research efficiency" described by Elmer K. Bolton, a highly successful research manager at the Du Pont Company from World War I until the early 1950s. Bolton, who presided over much of the company's research and development work in organic chemistry (including its famous discovery and development of nylon) would have had little to tell Edison about how to improve the efficiency of the Menlo Park Laboratory's research in 1879. [Elmer K. Bolton, "Research Efficiency," July 16, 1920, Records of E. I. du Pont de Nemours & Co., Central Research and Development Department, Accession 1784, Hagley Museum and Library, Wilmington, Delaware.]

Edison realized at roughly the same time that both the physical facilities and the technical capabilities of the Menlo Park laboratory were inadequate for the task he had undertaken. He therefore began a construction and equipment program that would make Menlo Park the most thoroughgoing facility in the world for electrical research. As he wrote at the time, he aimed to create a laboratory with:

> [A]ll the means to set up and test more deliberately every point of the Electric Light, so as to be able to meet and answer or obviate every objection before showing the light to the public or offering it for sale either in this country or in Europe. [Thomas A. Edison to Theodore Puskas, November 13, 1878.]

Edison also began to hire staff members who possessed the skills critical to the success of his venture, from scientifically trained men such as Upton to a variety of skilled tradesmen, including machinists and glass blowers.

Tight focus and thoroughgoing attention to detail thus changed the very nature of the Menlo Park Laboratory. No longer would it be a general, wide-ranging "invention factory." The laboratory was now a special-purpose facility, and as Edison and his staff rapidly developed the technology of electric lighting, Menlo Park began to lose its usefulness. One need not recount the entire history of Edison's development of the electric light to gain a sense of what was occurring at Menlo Park; that has been done well by Friedel and Israel. As they explain, by late 1880 Edison and his team had succeeded in the invention of a successful carbon-filament incandescent light bulb, an efficient dynamo to power such lights, and a distribution system that allowed the lights to be turned on and off at will. The Edison team had wired the Menlo Park complex—and Edison's and Upton's homes—into a mini-electric lighting system. Success, not failure, ultimately brought an end to Edison's "city upon a hill."

No Turning Back

With streams of capitalists, dignitaries, and city officials journeying to Menlo Park to see Edison's invention, the complex's power was now embodied in the physically visible spectacle of the electric lighting system; no longer was Menlo Park seen purely as the invention factory described by Edison in 1876, but rather as the site of the prototype lighting system of tomorrow. The *product* of the laboratory, not the *process* that had brought it into existence, now began to dominate Edison's and his associates' thinking. The very scale of the technology required for the development and implementation of Edison's system beyond that demonstrated at Menlo Park further demanded that Edison command facilities and skills greater than those accumulated at the rural Menlo Park complex.

Herein lies one of the most significant problems in both the management of industrial research and development and the management of successful technology-based startup companies: maintaining a truly creative atmosphere in a laboratory. The environment at Menlo Park that produced the major strides in Edison's electric lighting system was rapidly changing; from concerns about larger or broader research questions, the laboratory was quickly moving toward narrower, task-oriented R&D work.

The requirements for the implementation of a commercial-scale system demanded that Edison establish manufacturing operations for light bulbs, dynamos, and wire and conduit components. Key Menlo Park laboratory personnel moved into positions of management in those concerns, but no attempt was made to replace them with researchers of comparable skills who would have maintained the Menlo Park Laboratory as an invention factory. For a while, the Menlo Park Laboratory conducted the experimental work and necessary testing for the various manufacturing operations, but by the end of 1881, the various Edison factories had established their own testing laboratories and were doing their own experimental work. The creative juices of Menlo Park were being wrung from the laboratory and put into the tasks of commercializing the new technology. The Menlo Park Laboratory of 1882 was a mere vestige of what it had been in 1879 and 1880.

Precisely the same experience confounded the Du Pont Company in the 1930s, following the discovery of nylon by a group of fundamental researchers in the company's central research laboratory. Four years earlier the same research group had discovered neoprene synthetic rubber. At that time the research managers determined to transfer responsibility for neoprene's development and commercialization to the company's Organic Chemicals

Figure 103. *By the 1890s, several of the Menlo Park buildings were put to other uses. A family actually lived in the office, while the glass working shed became a chicken coop. Ultimately, most of the buildings fell apart or were dismantled.*

Department, which was well equipped and staffed for that purpose. But with nylon, those same managers decided that the central research laboratory should maintain control of the fiber's development because of its enormous potential and because its development was so highly dependent on the scientific expertise of the chemists in central research. As a consequence of this decision, researchers were pulled off existing projects to work on the development of nylon. The entire research agenda of the central research department was reoriented toward nylon. Although the fiber's development was a brilliant scientific, technical, and commercial success, the experience left Du Pont's central research department considerably different—and ultimately less productive scientifically—from what it had been in 1930. Other corporate R&D laboratories have gone through a similar process.

Some observers have credited the consistent success of AT&T Bell Laboratories in research to that institution's ability to transfer projects out of the research laboratory to other laboratories (e.g., to Western Electric, the manufacturing arm of AT&T) without pushing or allowing those involved in the discoveries to go into the organizations where the nitty-gritty of development and implementation occurs. Indeed, some observers fear that with the breakup of the Bell system and the new competitive pressures AT&T finds itself under, Bell Laboratories will succumb to the same problems; with greater need for quicker development of research leads, the character of the laboratory might change, just as the character of Menlo Park changed because of its success.

How AT&T chooses to develop its laboratories' research leads is clearly the choice of its managers. Edison had little choice but to allow Menlo Park, as it was in 1879 and 1880, to fall apart because his personal (and therefore his laboratory's) cash flow was tied almost entirely to the development of his electric lighting system. At the end of 1880, when he had demonstrated a successful prototype system at Menlo Park, Edison found himself in the same position many entrepreneurs in technology startup companies face today. Development and commercialization of a major new technology demand different organizational structures, and in startups it is difficult, if not impossible, to retain the old while building the new.

Edison ultimately responded to the situation by completely abandoning Menlo Park—essentially by 1882—and seeking to re-create the synergies of 1879 in an all-new and bigger laboratory at West Orange, New Jersey, once his work in electric lighting was finished. Edison built this new laboratory in 1887. By this time he had lost control of the electric light, outfoxed by the "robber barons," according to Matthew Josephson. But Edison had also lost whatever edge he personally had in the development of electric lighting technology, and at West Orange he moved into other technologies, some new and others familiar to him.

Menlo Park's Influence

What lasting influence did Edison's Menlo Park laboratory have on American industrial history and American history in general? Economic historian David Mowery has argued that 139 industrial research laboratories had been established in the United States prior to 1899 (almost all of them in manufacturing), but this number greatly exaggerates the extent to which industrial research and development laboratories were institutionalized in American business prior to the turn of the century. [David Mowery, "The Emergence and Growth of Industrial Research in American Manufacturing, 1899–1945," p. 51.] Mowery's figures include plant control and testing laboratories, most of which were small operations with few professional employees. Certainly none matched the size of the Menlo Park Laboratory at its zenith in 1879–81. Moreover, the missions of the majority of these laboratories were much narrower than the one established by Edison for his Menlo Park Laboratory in 1876. Much more research needs to be done by historians to evaluate the extent to which the fame of Menlo Park helped lead to the establishment of these other "research" laboratories. No other laboratory in the United States commanded as much prestige or was as well known as Edison's in the late 1870s and early 1880s. The founding director of the Du Pont Company's first R&D laboratory, Charles L. Reese, who would eventually head up a 1,000-person-strong R&D organization in the company and then move on to become president of the American Chemical Society and the American Society of

Chemical Engineers, cited Edison as his model for a research director.

The fame of Edison's Menlo Park Laboratory was twofold. First, the success of Edison the inventor led many other independent inventors to upgrade their "workshops" into "laboratories." Edison's rivals in telegraphy, telephony, and electric lighting aspired to gain fame comparable to Edison's and the resources to build laboratories similar to Menlo Park. None completely succeeded.

Second, while achieving fame in the popular culture of his time, Edison earned respect from the scientific community in the United States and Europe because of his achievements at Menlo Park. This had been, as noted above, one of Edison's goals for his Menlo Park venture. With George Barker's and Henry Draper's encouragement, Edison had been active with the American Association for the Advancement of Science and the more elite National Academy of Sciences. Both Edison and his scientist/assistant Upton prepared papers for the AAAS's annual meeting in Saratoga in 1879 on work related to electric lighting. Edison relied heavily on the established reputations of George Barker, Henry A. Rowland (the distinguished physicist at Johns Hopkins University), and two Princeton University physicists, Charles F. Brackett

Figure 104. *Modern corporate R&D centers, such as this General Electric Company facility, are often situated in isolated areas.*

and Charles A. Young, to verify his claims about the efficiency of his new dynamo electric machine. When the inventor sought to introduce his electric lighting system into Europe, his strategy included winning the support of well-respected scientists.

Edison actively cultivated science and scientists during the heyday of Menlo Park. He saw himself as more than an inventor; he considered himself to be a scientist as well. In late 1878, when embroiled in a controversy over the invention of the microphone with the Englishmen William Preece and David Hughes, he hastily drafted a reply to one of their missives, saying, "If prior publication is of no use, what protection has any scientific man. . . ." [Thomas A. Edison, October 7, 1878.] Two years later, when George Barker became an ardent supporter of his rival Hiram Maxim's electric lighting system (which Edison believed had been pirated from him), Edison wrote Henry Rowland, "There won't be much protection for a scientific man if his previous publications and exhibition counts for nothing." [Thomas A. Edison to Henry A. Rowland, December 13, 1880, Papers of Henry A. Rowland, Johns Hopkins University Archives.] Edison's self-image as a scientist was also

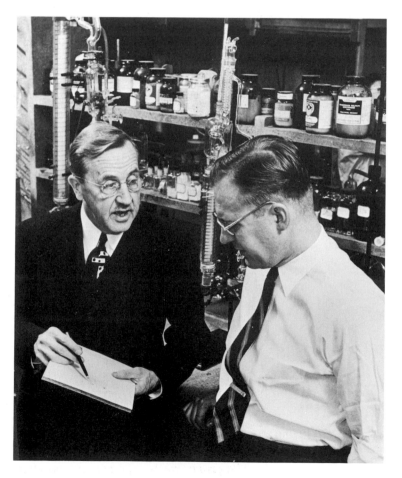

Figure 105. *Professor Roger Adams of the Chemistry Department at the University of Illinois, one of Du Pont's major consultants from 1927 to 1960, confers with a chemist at the company's experimental station near Wilmington, Delaware, in 1951. Adams served as dissertation advisor to many industrial chemists.*

Figure 106. *This facility of AT&T Bell Laboratories in New Jersey displays one type of layout common in R&D centers.*

manifested in 1880 when he became the founding publisher of *Science* magazine, the predecessor of the present AAAS-sponsored journal of the same name.

Yet some American scientists began to chafe at all the attention and praise accorded Edison and his Menlo Park Laboratory. Among the most outspoken was Henry Rowland. Perhaps he was jealous of Edison's Menlo Park Laboratory; certainly he wanted to build his own at Johns Hopkins University on a par with Edison's but for the pursuit of purer knowledge. Rowland was all too aware of what a writer for the *Electrical World* pointed out: "No university [has] the means of research . . . found in the shops of Thomas Edison at Menlo Park." [Quoted in Daniel Kevles, *The Physicists*, p. 139.] At an 1883 meeting of the AAAS, Rowland took up the cause of "pure science" and let go with a diatribe clearly aimed at Edison and those who praised him:

The proper course of one in my position is to consider what must be done to create a science of physics in this country, rather than to call telegraphs,

*electric lights, and such conveniences, by the name of science. . . .When
the average tone of the [scientific] society is low, when the highest honors
are given to the mediocre, when third-class men are held up as examples,
and when trifling inventions are magnified into scientific discoveries, then
the influence of such societies is prejudicial. . . .* ["A Plea for Pure Science,"
Physical Papers of Henry A. Rowland, p. 609.]

By 1884 Edison had already essentially parted ways with the scientists. He
had given up the publication of *Science*, and he had—at least temporarily—
given up invention. As he told a friend in 1883, "I'm going to be a business
man. I'm a regular contractor now for electric lighting plants, and I'm going
to take a long vacation in the matter of invention." [Quoted in Josephson,
Edison, pp. 269–70.]

The growth of pure science ideology in the United States—unquestion-
ably propelled in part by Edison and his Menlo Park Laboratory—established
a tension in the American scientific community that has persisted to this
very day. This tension is between academic scientists (who see their work
and their motives as "pure") and scientists who choose to work in industry
(and whose work is often regarded as "impure" and second class by their
academic counterparts). Such attitudes have often made it difficult to recruit
the highest caliber of scientists for careers in industrial research and development laboratories. However, those scientists who have left the academy to
enter the corporate R&D laboratory often have had experiences similar to those
of Menlo Park's Francis Upton. Not long after he joined Edison, Upton wrote
to his father, "I find my work very pleasant here and not much different from
the times I was a student. The strangest thing to me is the [salary] I get each
Saturday, for my labor does not seem like work, but like study and I enjoy it."
[March 2, 1879, quoted in Friedel and Israel, *Edison's Electric Light*, p. 148.]

But Upton, like Edison, found that the needs of Edison's electric lighting
venture pulled him out of the role of laboratory researcher and into the role
of plant manager in the light bulb factory. In many of today's R&D-driven
companies, like Du Pont, this change of roles is commonplace. Du Pont's
research managers have complained since the establishment of the company's
R&D laboratories (1902–03) that the constant movement of researchers out
of the laboratory and into plant management or sales has seriously under-
mined the necessary continuity of their laboratories' R&D programs. Yet such
transfers have materially benefited the commercial objectives of the company,
just as the transfers from Menlo Park to Edison's manufacturing operations
aided the commercialization of his lighting system.

The Menlo Park experience was not unique; its drama has been recast
in many arenas of American industrial history since Edison abandoned Menlo
Park. But certainly Edison's errand to that rural New Jersey hamlet gave us
an important glimpse of the way in which much invention and technological
development would be conducted in the 20th century.

Bibliography

This essay emerged from the research I conducted on Edison's relationship with the
American scientific community, which appeared as an article, "Edison and the Pure
Science Ideal in 19th-Century America," *Science* 207 (February 8, 1980): 612–617. In
preparing to write this essay, I carried out a broad reading in the new microfilm edi-

tion of the Thomas A. Edison Papers, especially the general correspondence files from 1876 to 1882 and selected other correspondence during this period. See Thomas E. Jeffrey, et al., eds., *Thomas A. Edison Papers: A Selective Microfilm Edition, Parts I and II (1850–1886)* (Frederick, Md.: University Publications of America, 1985 and 1987). I was struck as never before by the intensity of the Menlo Park experience and by the parallels I saw with many of the perennial issues confronting the management of research and development in the Du Pont Company during the 20th century, which John Kenly Smith, Jr., and I treat in *Science and Corporate Strategy: Du Pont R&D, 1902–1980* (New York: Cambridge University Press, 1988).

I have drawn from the standard secondary works on Edison, including Matthew Josephson, *Edison: A Biography* (New York: McGraw Hill, 1959) and Robert Friedel and Paul Israel, with Bernard S. Finn, *Edison's Electric Light: Biography of an Invention* (New Brunswick, N.J.: Rutgers University Press, 1986). My thinking about Edison and the Menlo Park experience has been influenced by the following works by Thomas P. Hughes: *Thomas Edison: Professional Inventor* (London: The Science Museum, 1976); "Edison's Method," in *Technology at the Turning Point*, ed. W. B. Pickett (San Francisco: San Francisco Press, 1977): 5–22; and "The Electrification of America: The Systems Builders," *Technology and Culture* 20 (April 1979): 124–161.

Literature on the history of industrial research and development has begun to grow into a significant body. Relevant works include Leonard S. Reich, *The Making of American Industrial Research: Science and Business at GE and Bell, 1876–1926* (New York: Cambridge University Press, 1986), George Wise, *Willis R. Whitney, General Electric, and the Origins of U.S. Industrial Research* (New York: Columbia University Press, 1985), Reese V. Jenkins, *Images and Enterprise: Technology and the American Photographic Industry* (Baltimore: The Johns Hopkins University Press, 1975), Stuart W. Leslie, *Boss Kettering* (New York: Columbia University Press, 1983), and David Mowery, "The Emergence and Growth of Industrial Research in American Manufacturing, 1899–1945," Ph.D. dissertation (Stanford University, 1981). My views on recent developments in corporate R&D stem from my work on Du Pont and from a wide array of business literature, much of it focused on the effect of the breakup of the Bell system on AT&T Bell Laboratories.

An important source on the working out of the pure science ideal in American history is Daniel J. Kevles, *The Physicists* (New York: Knopf, 1978). On Henry A. Rowland's views, see *Physical Papers of Henry A. Rowland* (Baltimore: The Johns Hopkins University Press, 1901).

Epilogue

THE MENLO PARK EXPERIENCE was central to several major historical transformations. No other place is more emblematic of the origins of modern sound and light technologies. No other laboratory is more symbolic of the evolution of modern research and development. Edison was instrumental in transformations that have come to epitomize modern society: haphazard to systematic innovation; print to visual/aural communications; individual to corporate research and development; experimental to theoretically based science; entrepreneurial to corporate capitalism; steam to electric power; artisan to industrial labor. Uneven and incomplete though these transformations may have been, the trends are irrefutable and Edison's Menlo Park Laboratory was a part of each of them.

It is in the relationships between the Menlo Park experience and its various contexts that we can learn specific ideas about invention and ingenuity. Edison combined his own basic research with the work of other inventors to focus on the systems that would bring inventions into the market place. He recognized the integral relationship between basic research and product innovation, and so should we. He defended lower profit rates in order to support basic research. He tied market and technological research, technical conceptions, visualization, fabrication, and testing into a tightly woven network. His nonbureaucratic approach has proven successful in other laboratory settings and its implications can be exploited by other inventors.

Finally, it has to be noted that Edison worked in a highly propitious economic atmosphere. America's producer-driven economy was expanding, technical innovations were seen everywhere, capital was available for investment, military spending did not drain resources, and population growth and incomes were high enough to generate market growth. Edison's laboratory was uniquely designed to take advantage of these external conditions. We should not expect a single research laboratory to be nearly as influential today, if only because there is more competition and more diversity.

Indeed, we should not be looking to make other Edisons; we should be looking at the constraints to the innovation process and we should be trying to overcome them wherever they appear, whether in technical knowledge, research organization, or market strategy. The Menlo Park experience illustrates how much a culture of ingenuity can influence inventiveness.

However, we need to be extremely careful when seeking to draw lessons from history. It would be a mistake to draw certain kinds of lessons. In the case of Edison and the Menlo Park experience, we should not conclude that technical creativity needs to be made more methodical and regularized. Edison attempted to approach problems systematically and to regularize the attention paid to research and follow-through. But creativity at Menlo Park developed out of the process in which laboratory assistants, machinists, and engineers responded in untraditional ways to the challenges they faced. Communication among the men was facilitated by the character of the community, by the openness of the second floor of the laboratory building, and by the

Figure 107. *The latest in laser technology was displayed at Menlo Park, Greenfield Village, in celebration of the 110th anniversary of Edison's successful incandescent light bulb experiment.*

widespread use of graphic material. The workers all felt that they could provide answers to the new questions that flowed in a constant stream from the on-going experiments relating to different projects. It was the interplay between disciplined analysis and free association that provided the stimulus for invention at Menlo Park.

Nor should we draw the conclusion that the Menlo Park team's successes derived from its seclusion. Employees clearly were not isolated from the workaday world that surrounded them; indeed, their experiences were but a variant of those of many other craft workers. Nor were they isolated from the results of their work. They might not have been able to anticipate all the results of the phonograph, telephone, or electrical lighting system. But they were not unmindful of the contexts in which their products would be used. The key was their ability to control public contacts and to work free from intrusions.

Unlike today's researchers, however, Edison's team did not have public and private agencies anticipating the environmental and economic impact of innovations. As a society we have become much more aware of the inter-relatedness of technical and social change and of the need to consciously make decisions about the benefits and costs of change. Indeed, the very fact that technical change is increasingly considered open to public discussion and political decision-making signifies one of the major differences between Edison's time and our own.

Further, we should not conclude that solely acquisitive individualism and free-market economics were the motors of innovation. Edison's corporate financial backers utilized monopoly profits to support research and government patents to control the flow of ideas. Personal ambitions for wealth and status were evident in Edison and his co-workers, but clearly other motivations were present as well. Workers willing to serve time learning a new occupation were investing in their own futures. These were risk-takers who knew that they had the opportunity for challenging work, not for instant gratification. The United States was a society that offered opportunity but did not guarantee success. Willingness to invest the time to fail and learn from failure is something that we collectively as well as individually could learn from Edison.

As to the historical Edison, it is evident that he was adept at identifying technical problems and in focusing research. The "Wizard" did not merely conjure up fantastic ideas; his propensity to think laterally and by analogy permitted him to reap the benefits of serendipity, as his initial thinking about the phonograph illustrates. He multiplied his own talent by subdividing the labor of research and experimentation and harnessing the inventive powers of others. The "Wizard" had the curiosity, envisioned the goal, and established the outlines of the puzzle; others provided many of the pieces required to solve it. The essays in this volume suggest that Edison's success stemmed from his methods of organizing and directing the work of others. They demonstrate that creativity is the product of a work process.

What Carlson and Gorman refer to as a "draghunt" is a powerful technique for those with sufficient resources to pursue the search wherever it might lead. The myth of Edison the tireless experimenter is hardly dispelled, but certainly must be qualified now to include the fact that he thoughtfully guided the research of others in particular directions with specific goals in mind. The myth of Edison the poor businessman, based largely on his career of making and losing fortunes, must be qualified to show that he was a shrewd and perceptive manager. We see glimpses of Edison the manager, a sufficiently savvy businessman to organize and administer an unprecedented undertaking. In fact, as Millard and Israel suggest, he was an innovative, as well as a traditional, entrepreneur. And Edison the denigrator of formal education and academic scientists must certainly be seen now as a person who acquired useful information from a variety of sources, scholars included, but insisted that the data be applicable to his purposes and not solely academic.

But merely revising these myths, probably more common in popular media and children's literature than in scholarly circles, is not the major point of these essays. Examining actual experience as a source of ideas about past lives is a tactic that is at the center of the museum experience. It is by surrounding ourselves with the tangible things—the tools, tables, instruments, inventions, and the buildings themselves at Menlo Park, Greenfield Village—that we begin to suspend our disbelief and imagine ourselves in the midst of the creative furor that enveloped Menlo Park, New Jersey.

By pairing this imaginative leap with the comparative perspective and critical edge provided by historical hindsight, the museum visitor and the professional historian come to share a sense of the process that fosters technical creativity. A shared sense of this kind is fundamental to the mission of museums, this museum in particular, for it brings historical perspectives to bear on contemporary issues. As citizens, we all—both museum visitors and professional historians—share a responsibility for future social decisions regarding technological research, development, and innovation. The social history of Menlo Park illuminated here becomes a part of the social history we are living today and will make tomorrow.

William S. Pretzer

Name Index

The Contributors

Bernard Carlson teaches the history of technology in the School of Engineering and the History Department at the University of Virginia. He is author of *Innovation as Social Process: Elihu Thomson and the Rise of the Electrical Industry, 1870–1900* (Cambridge: Cambridge University Press, 1991) as well as a number of articles on the invention of the telephone. With support from the Sloan Foundation, he is currently writing a biography of Edison's rival, Nikola Tesla.

Bernard S. Finn is curator of Electrical Collections at the National Museum of American History, Smithsonian Institution. He developed an exhibit entitled "Edison: Lighting a Revolution" (1979, extended to include lamp inventors in the late 20th century in 2000) and, with Joyce Bedi, a traveling photographic exhibit, "Edison after the Electric Light: The Challenge of Success" (1986, re-issued as "Edison after Forty" in 1996). He also collaborated with Robert Friedel and Paul Israel on *Edison's Electric Light: Biography of an Invention* (New Brunswick, N.J.: Rutgers University Press, 1986).

Michael E. Gorman is the chair of the Division of Technology, Culture and Communications in the Engineering School at the University of Virginia. He teaches courses on psychology of invention and ethics and science. His books include *Simulating Science: Error, Heuristics, and Mental Models* (Bloomington: Indiana University Press, 1992), *Transforming Nature* (Boston: Kluwer, 1998) and, with Matt Mehalik and Pat Werhane, *Ethical and Environmental Challenges to Engineering* (Upper Saddle River, N.J.: Prentice-Hall, 2000).

David A. Hounshell is David M. Roderick Professor of Technology and Social Change at Carnegie Mellon University. His *From the American System to Mass Production, 1800–1932* (Baltimore: Johns Hopkins University Press, 1984) received the 1987 Dexter Prize from the Society for the History of Technology. He is the author (with John Kenly Smith, Jr.) of *Science and Corporate Strategy: Du Pont R & D, 1902–1980* (New York: Cambridge University Press, 1988). He is working on a monograph on the history of industrial research and development in the United States.

Paul Israel is the managing editor of the book edition of the Thomas A. Edison Papers at Rutgers University and is the author of *Edison: A Life of Invention* (New York: John Wiley, 1998), which won the Dexter Prize from the Society of the History of Technology in 2000. He also has published *From Machine Shop to Industrial Laboratory: Telegraphy and the Changing Context of American Invention, 1830–1920* (Baltimore: Johns Hopkins University

Press, 1992) and is co-author with Robert Friedel of *Edison's Electric Light: Biography of an Invention* (New Brunswick, N.J.: Rutgers University Press, 1986).

Andre Millard teaches both economic history and the history of technology at the University of Alabama, Birmingham. He formerly served as an assistant editor on the Thomas A. Edison Papers Project at Rutgers University. He has published a history of Edison's business career, *Edison and the Business of Innovation* (Baltimore: Johns Hopkins University Press, 1990), and a history of sound recording, *America on Record* (Cambridge: Cambridge University Press, 1995).

Edward Jay Pershey is task force director for the development of a new Crawford Museum of Transportation and Industry for the Western Reserve Historical Society in Cleveland, Ohio. He joined WRHS in 1995 as director of education and curator of urban and industrial history. From 1987 to 1995 he was the founding director of the Tsongas Industrial History Center, an innovative hands-on museum education program in Lowell, Massachusetts. From 1981 to 1987, Dr. Pershey was supervisory museum curator at the Edison National Historic Site in West Orange, New Jersey.

William S. Pretzer is a curator at Henry Ford Museum & Greenfield Village. He has published articles on industrial design, technology education, and the history of labor and technology in the printing trade as well as an introduction to the reprint edition of volume one of Francis Jehl's *Menlo Park Reminiscences* (New York: Dover Books, 1990).

Harold K. Skramstad, Jr., president emeritus of the Henry Ford Museum & Greenfield Village, works in partnership with his wife, Susan, as a consultant to cultural organizations. They live in Las Cruces, New Mexico.

Photo Sources and Credits

Unless otherwise noted, all photographs and illustrated artifacts are from the collections of Henry Ford Museum & Greenfield Village.

The references are to figure numbers unless otherwise indicated.

AT&T Archives: 54, 66, 106.

Alexander Graham Bell Papers, Manuscript Division, Library of Congress: 68.

E. I. du Pont de Nemours & Co. Papers, Hagley Museum and Library: 105

Edison National Historic Site, National Park Service, U.S. Department of the Interior: Mina Miller portrait (p. 10), 33, 34, 52, 53, 73, 87, 88, 89, 90, 91, 92, 93, 94, 95, 96, 98, 99.

General Electric Research and Development Center: 104.

Harvard Historical Scientific Instruments Collection, Harvard University: 82.

Frank Leslie's Illustrated Weekly Newspaper (January 10, 1880): 36.

Prescott, George B., *Bell's Electric Speaking Telegraph* (New York: D. Appleton, 1884): 69, 77, 78, 79.

Scientific American (October 11, 1890): 45; (October 18, 1879): 48.

Urbanitzky, Alfred Ritter van, *Electricity in the Service of Man* (New York: Cassell and Company, 1886). Courtesy of Purdue University Library: 67, 81.